AND WHERE TO FIND IT

SCIENCE MATTERS

ALSO BY ROBERT M. HAZEN

Children of Prometheus with Margaret Hazen

The Breakthrough

The Music Men with Margaret Hazen

The Poetry of Geology

Wealth Inexhaustible with Margaret Hazen

North American Geology

ALSO BY JAMES TREFIL

Meditations at Sunset

Meditations at 10,000 Feet

A Scientist at the Seashore

The Moment of Creation

The Unexpected Vista

Are We Alone? with Robert T. Rood

From Atoms to Quarks

Living in Space

The Dark Side of the Universe

Space Time Infinity

Reading the Mind of God

The Dictionary of Cultural Literacy with E. D. Hirsch and
Joseph Kett

SCIENCE
MATTERS

ACHIEVING

SCIENTIFIC

LITERACY

ROBERT M. HAZEN
AND
JAMES TREFIL

DOUBLEDAY
NEW YORK   LONDON   TORONTO   SYDNEY   AUCKLAND

PUBLISHED BY DOUBLEDAY

a division of Bantam Doubleday Dell Publishing Group, Inc.

666 Fifth Avenue, New York, New York 10103

DOUBLEDAY and the portrayal of an anchor
with a dolphin are trademarks of Doubleday,
a division of Bantam Doubleday Dell
Publishing Group, Inc.

Typography and binding design by Viola Adams

Illustrations copyright © 1990 by Hardy

Library of Congress Cataloging-in-Publication Data

Hazen, Robert M., 1948–
Science matters: achieving scientific literacy / Robert M. Hazen
and James Trefil
p.   cm.
Includes index
1. Science—Popular works.   I. Trefil, James S., 1938–
II. Title.
Q162.H36   1991   90-3786
500-dc20   CIP

ISBN 0-385-24796-6
Copyright © 1991 by Robert M. Hazen and James Trefil
All Rights Reserved
Printed in the United States of America
January 1991

3   5   7   9   10   8   6   4

To Jeanne and Margee

CONTENTS

INTRODUCTION

SCIENTIFIC LITERACY: WHAT IT IS, WHY IT'S IMPORTANT, AND WHY WE DON'T HAVE IT

Sometime in the next few days you are going to pick up your newspaper and see a headline like "Genetically Engineered Tomatoes on Shelves" or "Japanese Take Lead in Superconductor Race." The stories following these headlines will be important. They will deal with issues that directly affect your life—issues about which you, as a citizen, will have to form an opinion if you are to take part in our country's political discourse. More and more, scientific and technological issues dominate national debate, from the greenhouse effect to the economic threat from foreign technology. Being able to understand these debates is becoming as important to you as being able to read. You must be scientifically literate.

Scientists and educators have not provided you with the background knowledge you need to cope with the world of the future. The aim of this book is to allow you to acquire that background— to fill in whatever blanks may have been left by your formal education. Our aim, in short, is to give you the information you need to become scientifically literate.

What Is Scientific Literacy?

For us, scientific literacy constitutes the knowledge you need to understand public issues. It is a mix of facts, vocabulary, concepts, history, and philosophy. It is not the specialized stuff of the experts, but the more general, less precise knowledge used in political discourse. If you can understand the news of the day as it relates to science, if you can take articles with headlines about genetic engineering and the ozone hole and put them in a meaningful context—in short, if you can treat news about science in the same way that you treat everything else that comes over your horizon, then as far as we are concerned you are scientifically literate.

This definition of scientific literacy is going to seem rather minimal, perhaps even totally inadequate, to some scholars. We feel very strongly that those who insist that everyone must understand science at a deep level are confusing two important but separate aspects of scientific knowledge. The fact of the matter is that *doing* science is clearly distinct from *using* science; scientific literacy concerns only the latter.

There is no need for the average citizen to be able to do what scientists do. You don't have to know how to calculate the trajectory of an artillery shell or sequence a section of DNA to understand the daily news, any more than you have to be able to design an airplane in order to fly in one.

But the fact that you don't have to know how to design an airplane doesn't change the fact that you live in a world where airplanes exist, and your world is different because of them. In the same way, advances in fields like microelectronics and molecular biology will affect your life in many ways, and you need to have enough background knowledge to understand how these changes are likely to occur and what their consequences are likely to be for you and your children. You must be able to put new advances into a context that will allow you to take part in the national debate about them.

Like cultural literacy, scientific literacy does not refer to detailed, specialized knowledge—the sort of things an expert would know. When you come across a term like "superconductor" in a newspaper article, it is enough to know that it refers to a material that conducts electricity without loss, that the main impediment to the widespread use of superconductors is that they operate only at very low temperatures, and that finding ways to remove this impediment is a major research goal in materials science today. You can be scientifically literate without knowing how a superconductor works at the atomic level, what the various species of superconductor are, or how one could go about fabricating a superconducting material.

Intense study of a particular field of science does not necessarily make one scientifically literate. Indeed, it has been our experience that working scientists are often illiterate outside their own field of professional expertise. For example, we recently asked a group of twenty-four physicists and geologists to explain to us the difference between DNA and RNA—a basic piece of information in the life sciences. We found only three who could do so, and all three of those did research in areas where this knowledge was useful. And although we haven't done an equivalent test on biologists—by asking them, for example, to explain the difference between a superconductor and a semiconductor—there is no doubt in our minds that we would find the same sort of discouraging result if we did. The fact of the matter is that the education of professional scientists is just as narrowly focused as the education of any other group of professionals, and scientists are just as likely to be ignorant of scientific matters as anyone else. You should keep this in mind the next time a Nobel laureate speaks *ex cathedra* on issues outside his or her own field of specialization.

Finally, one aspect of knowledge is sometimes lumped into scientific literacy but is actually quite different. You often see discussions of scientific literacy couched in terms of statements like "The average new employee has no idea how to use a computer" or "The average American is dependent on technology but can't even program a VCR to record when no one's home." These statements are probably true, and they undoubtedly reflect an unhappy state of affairs in American society. We would prefer, however, to talk of them in terms of *technological,* rather than scientific literacy.

The Scope of the Problem

At the 1987 commencement of Harvard University, a filmmaker carried a camera into the crowd of gowned graduates and, at random, posed a simple question: "Why is it hotter in summer than in winter?" The results, displayed graphically in the film *A Private Universe,* were that only two of the twenty-three students queried could answer the question correctly. Even allowing for the festive atmosphere of a graduation ceremony, this result doesn't give us much faith in the ability of America's most prestigious universities to turn out students who are in command of rudimentary facts about the physical world. An informal survey taken at our own university—where one can argue that teaching undergraduates enjoys a higher status than at some other institutions—shows results that are scarcely more encouraging. Fully half of the seniors who filled out our scientific literacy survey could not correctly answer the question "What is the difference between an atom and a molecule?"

These results are not minor blemishes on a sea of otherwise faultless academic performances. Every university in the country has the same dirty little secret: we are all turning out scientific illiterates, students incapable of understanding many of the important newspaper items published on the very day of their graduation.

The problem, of course, is not limited to universities. We hear over and over again about how poorly American high school and middle school students fare when compared to students in other developed countries on standardized tests. Scholars who make it their business to study such things estimate that fewer than 7 percent of American adults can be classed as scientifically literate. Even among college graduates (22 percent) and those with graduate degrees (26 percent), the number of Americans who are scientifically literate by the standards of these studies (which tend to be somewhat less demanding than our own) is not very high.

The numbers, then, tell the same story as the anecdotes. Americans as a whole simply have not been exposed to science sufficiently or in a way that communicates, the knowledge they need to have to cope with the life they will have to lead in the twenty-first century.

Why Scientific Literacy Is Important

Why be scientifically literate? A number of different arguments can be made to convince you it's important. We call them,

the argument from civics
the argument from aesthetics
the argument from intellectual connectedness

The first of these, the argument from civics, is essentially the one we have been using thus far. Every citizen will be faced with public issues whose discussion requires some scientific background, and therefore every citizen should have some level of scientific literacy. The threats to our system from a scientifically illiterate electorate are many, ranging from the danger of political demagoguery to the decay of the entire democratic process as vital decisions that affect everyone have to be made by an educated (but probably unelected) elite.

The argument from aesthetics is somewhat more amorphous, and is closely allied to arguments that are usually made to support liberal education in general. It goes like this: We live in a world that operates according to a few general laws of nature. Everything you do from the moment you get up to the moment you go to bed happens because of the working of one of these laws. This exceedingly beautiful and elegant view of the world is the crowning achievement of centuries of work by scientists. There is intellectual and aesthetic satisfaction to be gained from seeing the unity between a pot of water on a stove and the slow march of the continents, between the colors of the rainbow and the behavior of the fundamental constituents of matter. The scientifically illiterate person has been cut off from an enriching part of life, just as surely as a person who cannot read.

Finally, we come to the argument of intellectual coherence. It has become a commonplace to note that scientific findings often play a crucial role in setting the intellectual climate of an era. Copernicus's discovery of the heliocentric universe played an important role in sweeping away the old thinking of the Middle Ages and ushering in the Age of Enlightenment. Darwin's discovery of the principle of natural selection made the world seem less planned, less directed than it had been before; and in this century the work of Freud and the development of quantum mechanics have made it seem (at least superficially) less rational. In all of these cases, the general intellectual tenor of the times—what Germans call the *Zeitgeist*—was influenced by developments in science. How, the argument goes, can anyone hope to appreciate the deep underlying threads of intellectual life in his or her own time without understanding the science that goes with it?

What to Do

The beginning of a solution to America's problem with scientific literacy, both for those still in school and those whose formal education has been completed, lies in a simple statement:

If you expect someone to know something, you have to tell him or her what it is.

This principle is so obvious that it scarcely needs defending (although you'd be amazed at how often it is ignored within the halls of academe). It's obvious that if we want people to be able to understand issues involving genetic engineering, then we have to tell them what genetic engineering is, how DNA and RNA work, and how all living systems use the same genetic code. If we expect people to come to an intelligent decision on whether billions of dollars should be spent on a superconducting supercollider, we have to tell them what an accelerator is, what elementary particles are, and why scientists want to probe the basic structure of matter.

But this argument, as simple as it seems, runs counter to powerful institutional forces in the scientific community, particularly the academic community. To function as a citizen, you need to know a little bit about a lot of different sciences—a little biology, a little geology, a little physics, and so on. But universities (and, by extension, primary and secondary schools) are set up to teach one science at a time. Thus a fundamental mismatch exists between the kinds of knowledge educational institutions are equipped to impart and the kind of knowledge the citizen needs.

So scientists must define what parts of our craft are essential for the scientifically literate citizen and then put that knowledge together in a coherent package. For those still in school, this package can be delivered in new courses of study. For the great majority of Americans—those whom the educational system has already failed—this information has to be made available in other forms.

And that's where this book comes in.

About the Book

This book is dedicated to illustrating a statement that is one of academe's best-kept secrets: *The basic ideas underlying all science are simple.* In what follows, we present only the constellations of basic facts and concepts that you need to understand the scientific issues of the day, while providing a reading list for those whose interest has been piqued and who wish to pursue some of the points in more detail.

Science is organized around certain central concepts, certain pillars that support the entire structure. There are a limited number of such concepts (or "laws"), but they account for everything we see in the world around us. Since there are an infinite number of phenomena and only a few laws, the logical structure of science is analogous to a spider's web. Start anywhere on the web and work inward, and eventually you come to the same core. Understanding this core of knowledge, then, is what science is all about.

The organization of this book reflects the weblike organization of science. It is built around eighteen general principles—call them laws of nature, great ideas, or core concepts if you like. Some of them transcend the compartmentalized labels we like to put on things, for, like nature itself, these great ideas form a seamless web that binds all scientific knowledge together. We devote the first five chapters of the book to these concepts, which will reappear in the remainder. They are absolutely essential to understanding science. You can no more study genetics while ignoring the laws of chemistry than you can study language by learning nouns and ignoring verbs.

Once the basic concepts that anchor all science have been established, we move on to look at specific areas, which we have organized in the traditional triad of physical, earth, and life sciences. We organize each of these categories around another set of great ideas that are appropriate to that particular field. For example, in the earth sciences one of the great ideas has to do with the changes of the planet's surface features in response to heat generated in its deep interior. This particular concept ties together a great deal of what we know about the earth, but at the same time depends on the deeper overarching principles contained in the first five chapters. By the time you've gone through all eighteen great ideas, then, you will have not only a general notion of how the world works, but also the specific knowledge you need to understand how individual pieces of it (the earth's surface, for example, or a strand of DNA) operate.

The great ideas approach to science has another enormous advantage. While you are learning about issues that are in the news today, such as black holes or the greenhouse effect, you will be building an intellectual framework that will allow you to understand the issues of the future. To see the importance of this, consider that when we began writing this book in early 1989, the Strategic Defense Initiative ("Star Wars") was a much-debated aspect of public policy. A year later, thanks to the extraordinary events that took place in Central Europe, the topic had pretty much dropped from sight.

It is entirely possible that problems that loom large today—AIDS . . . the ozone hole—may seem insignificant by the turn of the century. But while we cannot predict *which* science-laden issues will dominate the headlines of the future, we know that some surely will. And since every future scientific advance will grow out of the ideas contained in this book, mastering them allows you to deal with not only today's problems, but tomorrow's as well.

There is a temptation, when presenting a subject as complex as the natural sciences, to present topics in a rigid, mathematical outline. We have tried to resist this temptation for a number of reasons. In the first place, it does not accurately reflect the way science is actually performed. Real science, like any human activity, tends to be a little messy around the edges. More important, the things you need to know to be scientifically literate tend to be a somewhat mixed bag. You need to know some facts, to be familiar with some general concepts, to know a little about how science works and how it comes to conclusions, and to know a little about scientists as people. All of these things may affect how you interpret the news of the day. So if you find that the book is something of a potpourri, don't be surprised. That's the way science—like everything else in life—is.

Finally, we hope that we can communicate to you another little-known fact about science: that it is just plain fun—not just "good for you," like some foul-tasting medicine. It grew out of observations of everyday experience by thousands of our ancestors, most of whom actually enjoyed what they were doing. Amid the collection of fact, history, logic, and policy questions that follows, we hope you occasionally catch a glimpse of two people who enjoy what they do—who occasionally thrill at the thought of the intellectual beauty of the universe we are privileged to live in and know.

SCIENCE
MATTERS

CHAPTER 1

KNOWING

Your life is filled with routine—you set your alarm clock at night, take a shower in the morning, brush your teeth after breakfast, pay your bills on time, and fasten your seat belt. With each of these actions and a hundred others every day you acknowledge the power of predictability. If you don't set the alarm you'll probably be late for work or school. If you don't take a shower you'll probably smell. If you don't fasten your seat belt and then get into a freeway accident you may die.

We all seek order to deal with life's uncertainties. We look for patterns to help us cope. Scientists do the same thing. They constantly examine nature, guided by one overarching principle:

The universe is regular and predictable.

The universe is not random. The sun comes up every morning, the stars sweep across the sky at night. The universe moves in regular, predictable ways. Human beings can grasp the regularities of the universe and can even uncover the basic, simple laws that produce them. We call this activity "science."

WAYS OF KNOWING

Science is one way of knowing about the world. The unspoken assumption behind the scientific endeavor is that general laws, discoverable by the human mind, exist and govern everything in the physical world. In its most advanced form, science is written in the language of mathematics, and therefore is not always easily accessible to the general public. But, like any other language, the language of science can be translated into simple English. When this is done, the beauty and simplicity of the great scientific laws can be shared by everyone.

Science is not the only way, nor always the best way, to gain an understanding of the world in which we find ourselves. Religion and philosophy help us come to grips with the meaning of life without the need for experimentation or mathematics, while art, music, and literature provide us with a kind of aesthetic, non-quantitative knowledge. You don't need calculus to tell you whether a symphony or a poem has meaning for you. Science complements these other ways of knowing, providing us with insights about a different aspect of the universe.

The Regularity of Nature

Our ancestors perceived the universe in ways that sometimes seem very strange to us. For all but the past few hundred years of human existence the universe was viewed by most people as a place without deep order or rules, governed by the whims of the gods or even by chance. By noting the daily movements of objec's in the sky, however, our ancestors got their first hints that some kind of order and regularity might exist in nature. The position of the sun, the phases of the moon, and the dominant constellations of stars cycled over the years, decades, and centuries with unerring regularity. Whatever governs its motion, the fact is that the sun does come up every morning.

Most historians of science point to the need for a reliable calendar to regulate agricultural activity as the impetus for learning about what we now call astronomy. Early astronomy provided information about when to plant crops and gave humans their first formal method of recording the passage of time. Stonehenge, the 4,000-year-old ring of stones in southern Britain, is perhaps the best known monument to the discovery of regularity and predictability in the world we inhabit. The great markers of Stonehenge point to the spots on the horizon where the sun rises at the solstices and equinoxes—the dates we still use to mark the beginnings of the seasons. The stones may even have been used to predict eclipses. The existence of Stonehenge, built by people without writing, bears silent testimony both to the regularity of nature and to the ability of the human mind to see behind immediate appearances and discover deeper meanings in events.

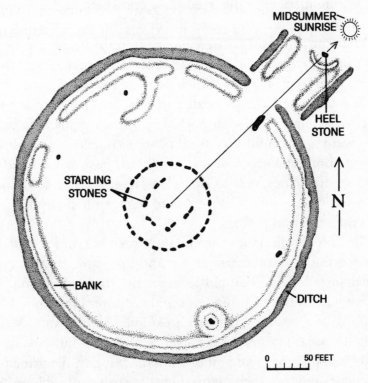

Stonehenge relied on the regular and predictable movements of sun, moon, and stars to serve its builders as a calendar. At the solstices and equinoxes, the light of the sun or moon aligns with the stones, and so documents the passage of time.

The Invention of Science

Astronomy was the first science. Throughout history some of the best minds produced by the human race have pondered the meaning of the celestial display. Most of the resulting theories shared a common property—they all assumed that in some way the earth was special, and that what happened in the heavens had

no relevance to phenomena on earth. In one important version of
the universe, for example, the stars and the planets turned eter-
nally on crystal spheres, and their motion had nothing to do with
mundane events like the fall of an apple in a orchard. People who
believed that the universe was built this way produced a large
body of accurate observations of the positions of heavenly bodies,
but astronomers were divorced from craftsmen and artisans who
were doing different things for the development of science.

While the astronomers were gazing into the heavens, other men
and women, equally ingenious, were trying to understand the way
things operated on earth. Their motivation was practical: they
studied the properties of heated metals because they wanted to
develop stronger alloys, they studied the flow of fluids because
they wanted to build canals, they experimented with different
combinations of ingredients to make better-tasting food and more
effective medicines, and so on. They never seemed to think that
the prosaic tasks in which they were engaged had anything to do
with the stars and planets.

The branch of science that finally broke out and forged a link
between the cerebral astronomers and the practical artisans was
"mechanics." This is an old term for the study of motion. Every
system, natural or man-made, contains matter in motion. Planets
orbit, blood circulates, chemicals explode, people walk. Mechanics is the superbly pragmatic science of pocket billiards and car
crashes, cannonballs and guided missiles. Today, the principles of
mechanics point to such useful things as stronger buildings, faster
cars, more exciting sports, and, as always, more sophisticated
weapons. But more important from the point of view of the birth
of modern science, the study of mechanics blazed the trail that
subsequent scientists have followed. While studying mechanics,
scientists developed and refined the scientific method, a technique that has given us so many new insights into the universe
we inhabit.

THE CLOCKWORK UNIVERSE

Modern science can be said to have started with the work of Isaac Newton (1642–1727) in England. According to Newton, the universe is something like a clock. In a clock, the external appearance—the slow sweeping of the hands—is a result of the motion of internal gears. In the same way, all of the natural phenomena we see in the world around us are the result of a few natural laws working beneath the surface of things. Newton demonstrated that:

One set of laws describes all motion.

For Newton, the key fact about motion was that it occurs in response to the action of one or more forces. The "gears" that connect forces and motion are Newton's three Laws of Motion, and they apply to everything that moves. Gases streaming out of an exploding star, a football thrown downfield, and blood cells in your arteries all move in compliance with these very simple, but very general, laws.

MOTION

Uniform Motion and Acceleration

If you're going to study something like motion, the first thing you have to do is decide what sorts of motion are found in nature. Scientists recognize only two kinds: uniform and accelerated. Everything in the universe is either in uniform motion or accelerating.

Any object that stands still or moves in a straight line at constant speed is in uniform motion. A book sitting on your desk, a car driving along an interstate with the cruise control set at 65 mph, and a spaceship traveling at 1,000 miles per second in deep space are all in uniform motion.

Acceleration is any change in motion and occurs when something speeds up, slows down, or changes direction. This definition may seem a little strange, because when you drive a car "acceleration" means speeding up—not slowing down or turning a corner. Physicists use a more general meaning for acceleration—but whatever the definition, it's something you feel in your gut. Flooring the gas pedal on your car, braking for a light, or rounding a bend all tend to move you around in your seat. And there's nothing subtle about acceleration—people don't ride roller coasters to experience uniform motion.

Newton's Laws and the Idea of Force

Isaac Newton, building on results from centuries of experiments on moving objects, wrote down a compact set of laws that describes the nature of all motion. That these laws apply to such an immense assortment of situations illustrates the power behind thinking of nature as regular and predictable. Newton's three Laws of Motion provide a cornerstone of physics and a model for what a science is supposed to be.

Newton's laws tell us how to predict the motion of a system just by knowing the forces that act on it. The three laws are stated separately, but they work together like separate gears that run a clock. Like all the fundamental laws that govern science, Newton's Laws of Motion may seem simple—almost simplistic. The deepest insights of the human mind often have this characteristic. Yet, as generations of physics students can testify, there is a subtlety and richness behind this apparent simplicity—how else could the laws describe everything from the orbits of Neptune's moons to the movement of exploding gases in your car's engine?

The First Law

Every body continues in its state of rest, or of uniform motion in a straight line, unless it is compelled to change that state by forces impressed upon it.

Newton has hidden two important concepts in this intuitively obvious statement. The first is inertia—the tendency of objects to continue doing what they're doing. A rolling ball keeps on rolling, a rotating planet keeps on rotating, a stationary book keeps on sitting.

The second concept is "force"—the thing that compels objects to change their state of motion (i.e., accelerate). Rolling balls can slow down if acted upon by a force. A book will move if pushed.

The point of Newton's First Law is that changes in motion do not happen spontaneously—there is always a reason for the change. A pencil falls, wind blows, popcorn pops. You encounter hundreds of examples every day. If an object accelerates, some kind of force must be acting. Behind every action verb is a force.

The First Law, by itself, says nothing about what forces are, what produces them, or how many different kinds there might be. Indeed, it took physicists more than two hundred years after Newton to discover the forces that hold atoms together, and we are still working to understand the force that cements the nucleus. Nevertheless, the First Law tells us what a force does when it acts, and, perhaps more important, it tells us how we can recognize situations in nature in which a force is present.

The Second Law

Force equals mass times acceleration

Newton's Second Law defines the exact relationship between an object's bulk, its acceleration, and the forces exerted on it. This is a commonsense sort of law that embodies two intuitively reasonable ideas. First, the Second Law says the greater the force, the greater the acceleration. The harder a pitcher throws, for example, the faster the ball travels. The more powerful your car engine, the better the pickup.

The second part of the law introduces the concept of mass, which is simply the amount of stuff being accelerated. Many of us use the words "mass" and "weight" interchangeably. That's not quite correct, because an object's weight depends on the local force of gravity (things weigh less on the Moon), but the mass depends only on how much stuff there is (how many atoms there are). Again, common sense prevails. Objects with lots of mass (refrigerators, boulders, football linemen) are a lot harder to move than objects with less mass (ice cubes, pebbles, quarterbacks).

The Second Law is quantitative—it can be written down as an equation ($F = ma$, if you really want to know). Numbers can be plugged into the equation to find out exactly how fast a spear, cannonball, or space ship of known mass will travel if it is acted upon by a known force.

In a typical mechanics problem, we know the mass of something (a billiard ball, for example, or a planet) and the force acting on it (the push of a cue stick or gravity). We then use Newton's Second Law and the branch of mathematics known as calculus to predict how the thing will move.

Why Newton Would Tell You to Wear a Seat Belt

Imagine yourself driving at 50 miles per hour along the freeway when another car forces you off the road. What happens if you smash into a tree head-on? Newton's Laws of Motion provide the answer.

You and the car have considerable inertia, which will be dealt with, one way or another, by the application of a force. The tree applies a force to the car, stopping it. In the absence of a seat belt, however, no force is applied to you, so you keep on moving. You are, in Newton's words "an object in a state of uniform motion," and you will therefore "continue in a state of uniform motion unless acted on by a force." The extent of your injuries will be determined by how that force is applied. Without a seat belt the driver and passengers will keep moving until they hit the steering wheel or the windshield.

Seat belts or self-inflating air bags act to slow you down at the same rate as the car. They apply a smaller force over a longer time, and that's a much safer method of applying the stopping force than hitting the steering wheel. The total change of motion with or without seat belts is exactly the same, but with seat belts the injury-causing force is not nearly so great.

The Third Law

To every action force there is an equal and opposite reaction force.

Even though this law is probably the most often quoted of the three, it is the least intuitive. It is obvious that a pitcher exerts a force on the ball, but less obvious that the ball pushes back on the pitcher's hand with an equal and opposite force. When you stand up, your shoes apply a force to the earth just as large as the force the earth's gravity exerts on you. When you try to open a screw-top bottle that is stuck, your left hand twists one way while the right hand is twisted the opposite way. You cannot touch your lover without feeling his or her touch in return.

The Third Law says that forces always come in equal and opposite pairs, but that the forces in the pairs act on (and therefore accelerate) different objects. You are pushing down on the chair in which you are sitting. The Third Law says that the chair is exerting an equal upward force on you. You really can learn Newton's Laws by the seat of your pants.

Newton's Third Law also explains how a rocket can fly in space, even when there's nothing to push against. It works like this: the rocket motor heats gases, which are accelerated out through the engine nozzle. The First Law tells us that in order to accelerate gas, we must exert a force on it. That force must, of course, be exerted on the gas by the ship. The Third Law then tells us that an equal and opposite force must be exerted by the gas on the ship. That's what makes the ship go. A rocket ship in space is similar to someone standing on roller skates and shooting a gun. Both recoil in one direction as they throw something out in the other.

GRAVITY

Newton's Laws tell us what happens when forces act on objects, but the laws tell us nothing about what those forces are. You'll discover several different forces in subsequent chapters, some well understood, like electricity and magnetism, some still

mysterious, like the so-called strong force. Newton himself described nature's most familiar force—gravity.

Before Newton, there was a kind of schizophrenia evident in
the way scientists thought about gravity. The force that held the
planets in their orbits (which we can call celestial gravity) was
held to be completely different from the force that makes things
fall to the center of the earth (terrestrial gravity). In the century
before Newton, different people made enormous progress in
studying these types of "gravity" separately.

Cannonballs

Terrestrial gravity was an obvious thing to study in an age when
cathedrals could collapse and cannonballs could sink ships. What
made the work of scientists in the seventeenth century different
from what had gone before was the appearance, for the first time,
of laboratory experiments—controlled studies of gravity's effect
on falling objects. The most famous of these experiments were
performed by the Italian scientist Galileo Galilei (1564–1642).
Galileo is best known for his trial on charges of suspicion of heresy for preaching the doctrine that the earth moves around the
sun (instead of vice versa), but in our view his most revolutionary
contribution to science was his demonstration that carefully run
experiments can yield profound insights into the nature of the
universe. He is, in fact, often called the "father of experimental
science."

Galileo studied terrestrial gravitation not by asking about the
nature of gravity, but by *observing* how objects behave when gravity acts on them. In particular, he did a series of experiments on
balls rolling down inclined planes (the purpose of the incline being, in his words, to "dilute" gravity enough so that he could
measure the time it took for the ball to roll with the primitive
clocks available to him). By meticulously measuring the time it
took the ball to travel various distances, he was able to find out
how the speed of the ball changed in transit. His bottom line:
terrestrial gravity causes all objects to accelerate the same
amount, regardless of their mass, and the rate of that acceleration

is constant. These simple observations allowed Galileo and his contemporaries to understand (and predict) things like the fall of a stone or the arc of a cannonball. They are the basic facts that tell you everything you need to know about how unsupported objects behave at the surface of our planet.

Ironically, Galileo probably never performed his most famous "experiment"—dropping two balls of different masses from the leaning Tower of Pisa to show that all objects fall at the same speed. Had he actually done the experiment, the resistance of the air might have caused the heavier objects to fall slightly faster than lighter ones, thereby disproving the very thesis he is famous for establishing!

Planets

It would be wrong to suppose that while Galileo was working out the effects of terrestrial gravity the astronomers were standing still. In fact, the German astronomer Johannes Kepler (1571–1630), using data on planetary motions assembled by the Danish astronomer Tycho Brahe (1546–1601), succeeded in discovering how the planets move in their orbits. He found, for example, that the orbits of all planets (including the earth) are elliptical—not circular, as everyone before had assumed. Like Galileo, he summarized his studies of planetary motion in concise statements, known as Kepler's Laws of Planetary Motion.

There are a number of important similarities in the methods used by Galileo and Kepler. Both men relied heavily on observational or experimental data. They were not, like many of their colleagues, armchair philosophers. If they wanted to know what the world was like, they actually went out and looked. Both men ended up by summarizing and codifying their results in a series of statements (or laws) written in mathematical form. These mathematical statements could be used by anyone to make predictions about the real world.

Kepler's Laws of Planetary Motion and Galileo's rules about falling bodies summarized the best scientific knowledge available in astronomy and physics, respectively, but they appeared to have nothing to do with each other. Each referred to a different sphere of reality. It took the genius of Isaac Newton to see that both men were, in fact, studying exactly the same thing.

The Apple and the Moon

According to Newton, he got his great idea while watching an apple fall in an orchard while he could see the moon in the sky. He knew the apple fell because a force acted on it (First Law), but it struck him that the force pulling on the apple might well extend all the way out to the moon and pull on that object too. It was this speculation, triggered by a simple everyday event, that led to the healing of the artificial distinction between the earthly and the heavenly, and that finally gave humanity both a new way to approach the world (science) and a new metaphor (the clock-work universe).

Newton knew that a dropped apple would fall straight down to earth under the influence of terrestrial gravity. Throw an apple straight out and it follows a curved path as gravity pulls it down. Throw the apple harder and it lands farther away. Throw it very hard indeed and it could even circle the earth. Once it makes one circuit, it will continue around and make another (ignoring air resistance), and will, in fact, continue to do so forever. But of course this is just what the moon (or any satellite) does. The force that constantly acts on the moon—that keeps pulling it into a curved path instead of the straight line the First Law says it should follow—is gravity, the same gravity that pulls down on the apple. With this insight, Newton abolished the centuries-old split between the earth and the heavens and showed that both were fit subjects for scientific study.

He went even further, deducing the exact mathematical formula for the gravitational force. Only three physical quantities determine gravitational force: the masses of the two objects, and the distance between them. He stated his result in what we know as Newton's Law of Universal Gravitation:

Between any two objects there is an attractive force
proportional to the product of the two masses divided by
the square of the distance between them.

This law has many interesting consequences. Obviously any large mass will exert a large gravitational force, but no special distinction is made between large masses and small ones. The earth pulls on the apple, but the apple also exerts a force on the earth. In fact, the two forces are the same size. We speak of apples falling to the ground because they are much less massive than the earth, and so undergo a much greater acceleration. But the earth also experiences an extremely small acceleration due to the force exerted by the apple. As the apple falls 15 feet to the ground, the earth "falls" a distance about the diameter of an atomic nucleus toward the apple.

The Law of Gravity tells us that every object in the universe is exerting a gravitational force on you right now. The earth exerts the biggest, but the person next to you exerts a force as well, as do the most distant star and galaxy. In practice, however, the massive sun and nearby moon are the only heavenly bodies that can exert a bigger force on you than familiar nearby objects like buildings. This simple fact is one of several reasons why scientists have a hard time taking astrology seriously.

The Clockwork Universe

With the Law of Universal Gravitation, Newton closed the circle on his work. He had the force—gravity—that operated everywhere, and he had the rules—the Laws of Motion—that governed the operation of all forces. Suddenly scientists saw the universe in a new way, ordered and predictable as never before. With Newton's equations and the language of mathematics, scientists could describe and predict the behavior of all kinds of systems. In the centuries following Newton's work, philosophers compared his vision of the universe to a clock. The visible phenomena in the world, like the hands of a clock, move in response to the actions of invisible gears—the natural laws. In the solar system the motions of the planets are governed by the Law of Universal Gravitation and the Laws of Motion. The planets tick along, as regular as a clock. For the Newtonians, in fact, the universe resembled a clock in other ways: once set in motion by God, the universe followed an inevitable course. The future was completely and comfortably predictable.

This is a wonderful vision, but like all scientific ideas it had to be tested. The most dramatic test of Newton's vision of the universe was made by his fellow Englishman Edmond Halley (1656–1742). Using Newton's Laws and historical records, Halley was able to work out the orbit of the comet that now bears his name and to predict its reappearance in the sky. When the comet was "recovered" on Christmas Day, 1758, the event powerfully underscored the idea of the clockwork universe. Not only could Newton's scheme explain things that were already known, it could make reliable predictions about events that had yet to occur.

Today, with the advent of quantum mechanics and the new field of chaos studies, scientist's ideas about the clockwork universe have changed. The universe is still, in the modern view, governed by simple laws, but these laws do not always allow us to make the kind of straightforward predictions about the future that Newton envisioned. Nevertheless, much of the Newtonian mindset survives in modern science.

THE SCIENTIFIC METHOD

Newton's development of the clockwork universe was the first, classic example of the scientific method in use. The method depends on a constant interplay of observation and theory; observations lead to new theories, which guide more experiments, which help to modify the existing theories.

In Newton's case, some of the observations and experiments were recorded by Galileo, others by Kepler. In each case, the cycle of observation, theory, test-against-new-observations was repeated until the investigators achieved a complete understanding of the phenomenon being studied. Newton, as we pointed out, incorporated these understandings into his sweeping theory of motion, and then his new theory was used to make many predictions like the projected reappearance of Halley's comet. Only after many such tests was the theory accepted by scientists.

The scientific method does not require researchers to be unbiased observers of nature. Scientists almost always have a theory in mind when they perform an experiment. But the method does require that scientists be willing to change their views about nature when the data demand it.

Newton provided a model for the development of modern science in many ways. He was the first to use the scientific method, and he was the first to show that scientific theories can develop by incorporation rather than revolution.

When Kepler published his laws of planetary motion, he swept aside the old ideas about the solar system. This was a revolutionary change—the old notions were seen to be wrong and were abandoned. When Newton published his work, however, he showed that all of Kepler's Laws could be derived from universal gravitation and the Laws of Motion. His work, then, incorporates Kepler's and expands upon it, but does not invalidate it. In the same way, Newton was able to derive Galileo's conclusions, incorporating them into the same theoretical framework that accommodated the description of the planets. This has proved to be a common occurrence in science. When Albert Einstein produced the Theory of General Relativity, our current best theory of gravitation, it incorporated Newton, Kepler, and Galileo, and when some future theoretical physicist produces the final unified field theory, it will likely incorporate Einstein.

Despite what you read in sensationalistic headlines, true revolutions are rare in mature sciences.

SCIENTISTS

The universe seems far too complex to comprehend all at once, so the classic scientific approach is to examine well-defined pieces of our surroundings, one at a time. The universe can be divided into an infinite number of "systems," which are nothing more than parcels of matter and energy. Each parcel, which can contain almost anything from a single spinning subatomic particle to an entire galaxy, is fair game for scientific study. Astronomers probe stars and the solar system. Chemists investigate systems containing carefully selected groups of atoms. Geologists study minerals or mountain ranges. Biologists examine complex systems called cells or ants or forests. Each system can be something you hold in your hand, like a rock, or it can be an integral part of something else, like your body's nervous system.

There are thousands of scientific subdisciplines, each with its own practitioners and jargon. These varied specialties differ primarily in the size and contents of the system under study. All systems, be they stars, bugs, or atoms, are governed by the same set of natural laws, but they are studied and described in very different ways.

About 350,000 Americans make their livings as scientists. Most of these women and men can be described with one of four broad labels: physicist, chemist, geologist, or biologist. Science is a seamless web of knowledge, but people like to create their niches. So each of the four main science branches (not to mention the hundreds of highly specialized "twigs") has developed its own distinctive style and organization.

Physicists study matter and energy, forces and motions—the concepts central to all science. Physicists take pleasure in pointing out that theirs is the most fundamental science, because all other fields, from chemistry to cosmology, mineralogy to molecular biology, depend on a few basic physical principles. Physicists are the generalists among scientists, and fields as far apart as molecular biology and field ecology have benefited from an influx of physicists over the years. Nevertheless, parts of physics have turned into the most abstract of the sciences. Physics conventions are replete with discussions of ten-dimensional space, quarks, and unified field theories. For some reason many physicists, particularly those in universities, seem to enjoy appearing sloppy and disheveled—always the ones without ties at faculty meetings. If you want to make a physicist happy, tell him you thought he was the plumber.

The American Institute of Physics, based in New York, represents more than 90,000 physical scientists, including astronomers, crystallographers, and geophysicists, who are members of ten affiliated societies. The largest of these groups, the American Physical Society, boasts about 40,000 hard-core physicists on its membership roles. These societies sponsor professional meetings, lobby for physics research and education, and publish prestigious research journals such as *The Physical Review* and *Physics Today.*

Chemists are pragmatists, studying atoms in combinations to discover new and useful chemicals. Most chemists, even those in academia, maintain close ties to industry; science and its applications are seldom far apart. Chemists hold more patents than any other kind of scientist, and they are frequently observed wearing business suits.

The American Chemical Society, headquartered in the nation's capital, represents both research chemists and chemical engineers. This blend of science and industry, unique among the major science societies, gives the ACS almost 140,000 members, making it the largest U.S. science society (surpassing even the interdisciplinary American Association for the Advancement of Science in total membership). The American Chemical Society sponsors meetings, supports chemical education, and publishes numerous books and journals, including the weekly *Chemical and Engineering News.* As a lobbying organization, the ACS must walk a fine line between environmentalists and major chemical corporations, both of whom are represented among the membership.

Geologists are a different breed. They frequently lecture in worn jeans and sturdy boots, seemingly ready to hike miles in the wilderness carrying rocks on their backs. Geology attracts men and women who love the outdoors and like to get their hands dirty. In practice, not all geology is rugged. The earth sciences employ much of the sophisticated lab hardware of chemistry and physics to decipher the nature and origin of rocks and minerals, oceans and atmospheres.

Most U.S. geologists belong to one of two earth science societies, each of about 20,000 members. The Geological Society of America, headquartered in Boulder, Colorado, caters especially to field geologists. The Washington-based American Geophysical Union encompasses a broad range of research, from planetary geology and physics to meteorology and oceanography. Both societies are active in international projects because geology is a global science, requiring global cooperation.

Biologists study life, the most complex systems of all. There are so many levels at which to study living things—molecules, cells, organisms, ecosystems—that biologists are a rather fragmented group. The concerns of zoologists working at zoos are quite different from those of industrial genetic engineers or hospital medical researchers. Consequently, there is no central American biological society, nor is there a strong lobbying presence in Washington, D.C.

Most biology funding, and hence most research, is related to human health. Ties between biological researchers and medical centers, such as the National Institutes of Health, are especially strong, and influence everything from lab design (biology labs often look and smell like hospitals) to dress codes (biologists are the *only* group of scientists who routinely wear white lab coats).

FRONTIERS: CHAOS

Newton's laws of motion and gravity were published more than 300 years ago and his so-called "classical" mechanics is now a well established part of any freshman physics course. But although the regularity and predictability of the universe has become an ingrained assumption of our science, recent studies of complex systems like your heart and the weather are making scientists rethink the meaning of predictability. This new development goes under the flashy title of "chaos."

Many day-to-day systems are predictable in the conventional sense. Automobiles, tennis balls, and grandfather clocks act pretty much the way we expect them to. If you drop a tennis ball from waist high, it hits the ground at one speed: drop it from a little higher up, it hits the ground at a slightly higher speed. A falling tennis ball is a conventional Newtonian system.

There are, however, systems in nature that don't display this sort of pleasing regularity. Open the tap on your faucet and you get a small, slow stream. Open it a bit more and you are apt to get a rushing, turbulent flow. In the jargon of physics, the behavior of tap water is extremely sensitive to the initial conditions. A system with this property is said to be chaotic; turbulent streams, growing snowflakes, your heart rhythms, and many other systems are chaotic.

The point about chaotic systems is that you can never measure the initial conditions of a system accurately enough to allow you to predict its behavior for all future time. Although your predictions and the real system may be close to each other for a while, eventually they will diverge. The inevitable errors inherent in any measurement, coupled with the extreme sensitivity of chaotic systems to initial conditions, means that for all practical purposes they are unpredictable (although they are perfectly predictable if the initial conditions are specified with mathematical precision).

The weather provides the most familiar example of a chaotic system. Meteorologists make thousands upon thousands of measurements of wind speed, air temperature, and barometric pressure in their efforts to predict the weather. They do pretty well with 24- and 48-hour forecasts, and sometimes they even get the seven-day predictions right. But no matter how fancy the measurements and the computer simulations, there is no way to guess what the weather will be a year from now. The chaotic nature of atmospheric motion can be graphically represented by something called the "butterfly effect," which is the statement that in a chaotic system an effect as small as a butterfly's flapping its wings in Singapore may eventually make it rain in Texas.

Today, the existence of chaotic systems is accepted by scientists, who now ask which systems are chaotic, how they behave, and how our newly won knowledge of that behavior can be utilized. Can we, for example, produce accurate monthly forecasts of the weather or the stock market?

CHAPTER 2

ENERGY

Aficionados think the old wooden roller coasters are still the best. If you've ever ridden one you're not likely to forget the experience. The adventure begins calmly enough, as you lie back in your cushioned seat enjoying the gradual climb to the ride's highest point. The steady clack-clack-clack of straining gears belies the wildness to follow. In the best of roller coaster tradition the car comes almost to a stop, poised at the brink, before gravity takes over. Then comes the plunge.

Faster and faster you go, 100 feet of free fall before the car hits bottom and zooms to the next height, only slightly shorter than the first. For a second time you almost come to a stop, and then another precipitous drop. Now the speeding ride takes off for a series of twists and turns, flinging you around tight loops and over violent bumps leading to one last mighty hill. The journey lasts only a couple of minutes, but you emerge, wobbly-legged, with a high that can last for hours.

Roller coasters are a microcosm of the universe. You were probably a bit too preoccupied to think about it the last time you rode one, but as you climbed and zoomed down those hills, hitting all the fancy loops and bumps, you demonstrated the basic laws of how energy behaves. Everything you do or see requires energy, and that energy always follows two basic rules:

Energy is conserved.
and
Energy always goes from more useful to less useful forms.

These two laws suggest good news and bad news. The first rule says that energy, the ability to perform useful work, comes in many different forms, and these forms are interchangeable. You can shift energy from one form to another the way you shift money between bank accounts. But just as transferring money neither adds to nor detracts from your wealth, changing energy from one form to another does not change the total amount of energy available. Energy cannot be created or destroyed, so the total energy of an isolated system stays the same. This is known as the First Law of Thermodynamics. It's the good news.

The bad news about energy is that its conversion always leads from more concentrated (i.e., more useful) to less concentrated (less useful) forms. When you burn coal or gas, the laws of nature require that some of the high-grade energy in the fuel must be dispersed as heat in the atmosphere, where it cannot be recovered to perform useful work. This is called the Second Law of Thermodynamics.

These two Laws of Thermodynamics govern all processes that involve heat and other forms of energy.

WORK, ENERGY, AND POWER

Thermodynamics relies on three concepts: work, energy, and power. Each of these words has an everyday meaning, but scientists use them in slightly different, specialized ways.

As far as physicists are concerned, work is done every time a force is used to move something. The amount of work done depends on how much force is used and how far the object moves (work equals force times distance). When you lift groceries, open a door, or throw a ball, you apply a force over a distance and move an object. You do work. The greater the force applied or the longer the distance moved, the more work is done. If you try to move something and fail (try to lift a car, for example, or push a wall) you haven't done any work. You may have generated a force, but it was not applied over a distance. No matter how much effort you expend, if nothing moves, no work is done. Physicists can thus actually prove that bricklayers do more work than lawyers.

Energy is the ability to do work—the ability to exert a force.

Power is a measure of how quickly work is done: work done divided by the time it takes to do it. If you run up a flight of stairs instead of walking, you require more power, even though the total amount of work done is the same in both cases. In sports, the player who can generate power at the highest rates throws farther, hits harder, and runs faster.

TYPES OF ENERGY

Energy comes in many forms. It can be converted from one form into another, but all the forms have one thing in common: they involve a system that is capable of exerting a force.

Potential Energy

A boulder perched at the edge of a cliff has the potential to do work. If it is allowed to fall, it will exert a force as it creates a depression in the earth below. It must therefore possess energy while it's sitting quietly at the top. We call this potential energy, where the term "potential" signifies that the system could do work, but isn't doing so at the moment. Think of the boulder as storing energy against future use.

If you dam a river, the water high in the reservoir has gravitational potential energy. We store the energy until we want to use it, at which time the water is allowed to fall and do work. This is a common technique for generating electrical power, particularly in the northwestern United States.

There are many different kinds of potential energy. Stretch a rubber band or compress a spring and they are ready to snap back, exerting a force over a distance as they do so. We say the rubber band has elastic potential energy. Coal, gasoline, and other combustibles store chemical potential energy, which is released to do work when they are burned. Energy can also be stored in magnets, in batteries, in the surface tension of drumheads and soap films, and in many other systems in nature.

Kinetic Energy

Anything that moves possesses energy. A fastball headed toward home plate, a rotating waterwheel, a speeding car, or a falling leaf can do work. You see this anytime a moving object stops. A moving baseball exerts a force on the catcher's mitt, compressing the material and leaving an indentation. It exerts a force over the distance of the compression. A speeding car that stops suddenly (by hitting a tree for example) exerts a force that makes the tree move. A rock slide or avalanche can obliterate an entire town. By virtue of the fact that it is moving, each of these objects can do work, and therefore possesses kinetic energy.

Every atom in a material is in motion, either moving freely (as in a gas) or vibrating (as in a solid). The kinetic energy of all these atoms is related to the phenomenon we call heat. The more vigorously atoms move, the greater the heat energy in the material of which they are part. This way of explaining the nature of heat says, in effect, that heat is simply a special kind of kinetic energy—the kind associated with atoms in motion.

The recognition that heat is a form of energy was one of the great insights of nineteenth-century science, and is the foundation stone of thermodynamics. Although the notion seems simple to us (particularly when we picture heat as vigorously moving atoms), it is far from obvious that there is a connection between a pot of boiling water and a boulder perched on a hilltop.

Other Kinds of Energy

When electrical current flows in a wire, electrons are moving. The kinetic energy of those electrons is one sort of electrical energy—one that lights your lamps, runs your stereo, and perhaps even cooks your food.

Sound is a special kind of kinetic energy caused by regular patterns of atomic movement. We hear sounds because our eardrums vibrate in response to moving air molecules.

Visible light carries with it energy that is related to electricity and magnetism. Other forms of radiation, such as radio waves and X rays, carry the same kind of energy.

In the twentieth century, a new category was added to the roster of known types of potential energy—mass. The equivalence of mass and energy is embodied in science (and in folklore) by Einstein's famous equation: $E = mc^2$. The equation says that mass can be converted into other forms of energy (and vice versa).

GOOD NEWS—THE FIRST LAW

One of the most important features of energy is that it easily shifts from one form to another. When you ride a bicycle, chemical energy in your cells changes into the motion of your legs and, ultimately, into kinetic energy as the bike moves. When you climb a hill some of your energy is converted to gravitational potential, and when you coast down the other side of the hill, that potential energy is converted back into kinetic energy as you speed up.

No matter how often energy is converted from one form to another, there is one simple rule that always applies. The total energy—the sum of energy in all its forms—is the same after the conversion as it was before. Energy cannot be created or destroyed, only shifted from one type to another. We say that the total energy is "conserved," and call this First Law of Thermodynamics the "conservation of energy." Conservation of energy is one of the deepest and most widely applicable principles in the sciences, and operates everywhere from the largest supercluster of galaxies to the cells in your body to the smallest, most fundamental pieces of matter that we know about, the quarks. It is one of the great integrating principles of science.

The First Law of Thermodynamics is the joy of amusement park owners and the bane of dieters. It tells us that the higher the roller coaster climbs, the faster we'll go. It also tells us that the chemical energy or calories we take in as food must either be used to do work (exercise) or stored (as fat). The spate of new diet strategies that come out every year are a testimony to the power of the First Law and to the futility of trying to avoid it.

An easy way to picture the operation of the First Law is to think about that roller coaster ride. You start at the bottom and powerful motors gradually pull the coaster up a steep incline to its highest point. During this phase of the ride electrical energy is converted to gravitational potential energy. At the top, gravity takes over and the car starts down. The car exchanges its gravitational potential for kinetic energy as it descends, moving faster and faster until it gets to the bottom. At this point in the ride, all of the original gravitational potential energy has been transformed to kinetic energy, and the car is moving as fast as it can go. As it starts back up the hill, it slows until it almost comes to rest at the next summit. And so it goes, potential and kinetic energy passing back and forth as the car goes up and down.

In the absence of friction, this transfer between potential and kinetic energy would go on forever. At each location on the track the total energy of the car—the sum of potential and kinetic energies—is exactly the same. Energy is neither created nor destroyed. The process is like transferring $100 back and forth between two bank accounts.

In the real world, there are other shifts of energy—the heat energy of wheel friction, the sound energy of straining wood and steel, the vibrational energy of the wooden framework and ground. Energy transferred into these categories does not flow back into either potential or kinetic energy; it is lost to the environment. Thus, on each hill the amount of potential energy is a little less than it was on the preceding one (that's why the second hill on a roller coaster is always lower than the first). The effect of friction is analogous to taking a dime out as a service charge every time you shift your $100 from one account to another. The dime doesn't disappear, of course—it goes to the banking company—but eventually the amount of money in your account is reduced to zero. In the same way, a roller coaster will eventually come to a halt, even if no brakes are applied. The energy hasn't disappeared, but it has shifted from the roller coaster to other, less exciting accounts.

Living systems operate a lot like roller coasters. We take in chemical potential energy in the form of food, which is modified and stored in our cells. That chemical potential energy is constantly used to keep us warm, drive our muscles, and perform work. You can't do anything without energy. Even thinking and sleeping consume it. So life is a constant struggle to gather and use energy.

HEAT

Moving Heat

Not only can energy be converted from one form to another, it can also be moved from one place to another. This is particularly true for heat energy. Heat flows from the burner on your stove into your food, for example, and from a drink into an ice cube. The transfer of heat plays a crucial role in many systems in nature, and is involved in effects as different as the movement of continents on the earth and the growth of plants. Heat moves from one place to another in three different ways: conduction, convection, and radiation. We experience all of these effects every day of our lives.

Put a metal spoon in a bowl of hot soup and the handle becomes hot in a matter of seconds—metal conducts the heat. If you had a very powerful microscope that allowed you to see inside the spoon, you would see its atoms near the soup moving very fast because they had suffered collisions with fast-moving water molecules in the soup. Those fast-moving atoms in the metal then collide with slower-moving molecules farther up the spoon, which collide with atoms still farther up and so on. Eventually, the atoms at the very end of the handle are moving fast and we say the spoon is hot. *Conduction* thus relies on the transfer of heat energy through the motion and collisions of individual atoms.

Water boiling on a stove and hot air rising from the asphalt on a parking lot in summertime are both examples of heat transfer by *convection*. Think about the water in the pot. Heat from the stove causes the molecules of water in the bottom layer of the pan to move faster. Some heat travels up through the water by conduction, but the process is too slow and cumbersome to move all the energy that is pouring in. The water at the bottom heats up, expands, and rises, to be replaced by cold water from the upper regions of the pot. When this heated water gets to the top it cools off and sinks, to be replaced by newly heated water from the bottom. The cycle of heating and cooling goes on, creating what is known as a convection cell. Convection depends on heat being carried from one place to another by the bulk motion of warmed materials, rather than through collisions between individual atoms.

Hold your hand out toward a fire and you feel warmth, even though neither convection nor conduction is operating. Heat reaches you by *radiation*. In this case, infrared radiation (an invisible cousin of ordinary light) travels from the fire to your hand, carrying energy in the process. Every object in the universe gives off heat by radiation. Indeed, for something like a satellite or a star in the vacuum of space, radiation is the *only* way that heat can be given off.

Heat, Temperature, and Absolute Zero

The words "temperature" and "heat" are often used interchangeably, but scientists think of the two terms in quite distinct ways. Heat refers to the total amount of atomic kinetic and potential energy in a material. Two gallons of ice water hold twice as much heat energy as one gallon. Temperature, on the other hand, is a relative term. Two objects are at the same temperature if no heat flows between them. Put a pound of metal into a gallon of ice water and they will soon be at the same temperature—32° Fahrenheit. But the metal and ice water do not hold the same amount of heat energy, because it takes more energy to make the atoms of water vibrate.

The time and temperature display at your local bank probably gives temperatures in degrees Fahrenheit and degrees Celsius, two scales in common use. The choice of which scale to use is arbitrary—there is no "correct" way to report temperature. All you have to do is pick two easily reproducible temperature reference points, assign a number to each, and then split the gap between the numbers into convenient intervals that you call "degrees." The Celsius scale, for example, uses the freezing and boiling points of water as its two reference points, calling the former zero, the latter 100, and defining a degree Celsius as one one hundredth of the interval between the two. The Fahrenheit scale works the same way, with zero signifying the coldest temperature Daniel Fahrenheit could produce in his laboratory back in 1717 and 100 being his best determination of human body temperature. Today, the Fahrenheit scale is defined in terms of the freezing and boiling points of water (32° and 212° respectively), and human body temperature is the more familiar 98.6° F.

Insofar as there is a "scientific" scale, however, it is what scientists call the Kelvin scale, named after William Thomson, Lord Kelvin (1824–1907), one of the founders of thermodynamics. The degree in the Kelvin scale is the same as a degree Celsius, but, unlike other scales, it is anchored by the only temperature that refers to a fundamental physical process.

The zero in the Kelvin scale is taken to be absolute zero–the lowest temperature attainable by any natural system. In the nineteenth century, absolute zero was pictured as the temperature at which all atomic motion stopped—at which everything just froze. Today, the laws of quantum mechanics have changed this picture slightly, and we define absolute zero as the temperature at which no more heat can be extracted from a system. Either way, absolute zero is cold. It measures $-273.16°$ C. (or $-453°$ F.). On the Kelvin scale, water freezes at 273.16 K., room temperature is about 300 K., and a wood fire ignites at about 650 K.

BAD NEWS—THE SECOND LAW

The First Law tells us that energy can be converted from one form to another and that the total amount of energy in a closed system is fixed, but it says nothing about whether a particular store of energy can be used to do anything useful. There is, for example, a great deal of energy stored in the vibrational energy of water in the ocean. According to the First Law, that energy could, in principle, be used to power a ship. The fact that no one has devised a ship to tap this reservoir of energy is a consequence of the Second Law of Thermodynamics. This Law places limitations on the ways that heat can be converted to useful work, and, in the process, produces a gloomy picture of the evolution of the universe.

Like other fundamental laws, the Second Law is deceptively simple, but belies a great deal of depth. Unlike most of the other laws we will encounter, however, the Second Law can be stated in ways that appear at first to be quite different. In fact, they are all logically equivalent—any one statement implies (and is implied by) the others. First statement:

Heat energy always flows spontaneously from hot to cold

If you put an ice cube on the table, heat will flow into it from the surrounding air. The ice cube doesn't get successively colder as heat flows out of it into the air. This commonplace experience is one illustration of the Second Law.

It is important to realize that the Second Law does not say that heat never flows from cold to hot. When you make ice in a refrigerator, that's exactly what happens. What the Second Law says is that if you want heat to flow "against the grain," you have to put energy into the system. In a refrigerator, energy comes into the system through the electric power cord. In fact, we like to say that there is another statement of the Second Law, namely:

A refrigerator won't work unless it's plugged in.

Second statement:

> *It is impossible to build an engine that operates on a cycle whose only effect is to convert heat into an equivalent amount of work.*

An engine is a device that takes energy stored in some form and produces useful work. The engine in your car, for example, takes chemical energy stored in gasoline and produces kinetic energy to make the car move. The Second Law says that no engine can operate at 100 percent efficiency.

At one level this isn't too surprising, especially for real engines. Every real machine must have moving parts, and every time parts move, some energy is lost as frictional heat. Heat always flows from hot to cold, so the frictional heat dissipates; it flows out to the machine's surroundings and that lost energy can do no further work through the machine. When you drive your car, some of the potential energy of the gasoline is lost as engine heat, as tire wear and friction, or even as sound energy as the car speeds along.

But even if there were no friction in an engine, the Second Law tells us that the efficiency with which heat can be converted to useful work must be less than 100 percent; some energy must always be dissipated into the environment as waste heat. The reason for this is that every engine, no matter what its design, must operate on a cycle, and must always be returned to its original position so that it can start its cycle over again. In your car, for example, the gasoline-air mixture that explodes in your cylinder drives the piston down. This motion, through a number of intermediate steps, eventually becomes the force that turns the wheels of the car, moving it forward. In principle (but not in practice) this operation could be carried out with 100 percent efficiency—all the stored energy of the fuel converted into the kinetic energy of the car. The important point, however, is that at the end of this operation, the piston is at the bottom of the cylinder. It has to be returned to the top before it can make the car move farther. To return the piston to the top, you have to cool the cylinder down so that the air you are compressing doesn't finish the cycle at a higher temperature than it had at the beginning. In your car, this heat (as well as some of that generated by the explosion) is carried away from the cylinder by the cooling system and transferred to the atmosphere by the radiator.

According to the Second Law, every engine, no matter how cleverly designed, must take some of its original store of energy and transfer it to a lower-temperature reservoir in order to return the engine to its original position. In most cases, this reservoir is the atmosphere or the ocean. In that reservoir the energy becomes unusable, because to tap it we would have to build an engine that could deposit its waste heat in some still cooler reservoir that doesn't exist. To argue otherwise is equivalent to saying you can make a refrigerator that will work when it isn't plugged in.

The Second Law leads to the intuitively reasonable notion that energy sources appear in a hierarchy, with high-grade sources producing very high temperatures and dumping heat to lower-temperature reservoirs, heat from those reservoirs dumping heat to those at a still lower temperature until, at the end of the chain, the accumulated waste heat ends up in the environment.

This idea has important technological consequences. The Second Law tells us, for instance, that if we burn coal with perfect efficiency, only slightly more than a third of the energy locked in the coal can ultimately appear as electricity in your home. The rest of the coal's chemical energy must go into the atmosphere— a fact that explains the presence of giant cooling towers at large generator sites. The true efficiency of a generating plant must be even smaller than this to take into account the effects of things like friction. To give credit to our engineering colleagues, we should point out that the efficiency of modern generators is within a few percentage points of the limit allowed by the Second Law.

Every few years for the past two centuries enthusiastic inventors have come forth with claims of perpetual motion machines— miracle devices that run forever with no additional energy input. So far none of the machines has worked. (One of our colleagues boasts that he keeps a complete collection of successful perpetual motion devices in his desk drawer.) New claims are met with universal skepticism by the scientific community, but not because scientists are too conservative or unwilling to listen to new ideas. Such machines would violate the Second Law, and would be equivalent to refrigerators that work when they're not plugged in. Third Statement:

> *The amount of disorder in any isolated system cannot decrease with time.*

If you placed 100 black marbles in the bottom of a jar, put a layer of 100 red marbles above them, then put a layer of 100 green marbles on top and shook the jar vigorously, you know what would result. In a short time black, red, and green marbles would be all mixed together. You could keep shaking for a million years and still never come close to duplicating the original orderly configuration.

This example illustrates an important point about nature. Isolated systems naturally tend to move from order to disorder. Put another way, time has a definite direction, with disordered systems occurring later than ordered ones. This idea is deeply ingrained since your life is full of examples of the Second Law. Just play your home movies backward and see how many things look wrong. It's easier to make an omelet than to unmake it, easier to scratch the side of your car than to paint it, easier to mess up your room than to clean it, and so on. All of these examples express the same idea of the directionality of time in nature.

"Entropy" is the scientist's measure of a system's randomness or disorder. The third statement of the Second Law says that in any closed system entropy either increases or (at best) remains the same over time. Disorder never decreases. Having said this, however, we have to admit that some systems can become more ordered—but at the expense of making things more disordered someplace else. You can vacuum your room, but you also have to pay the electric bill. When you make an ice cube in your refrigerator the water becomes more ordered as it freezes. The refrigerator, however, is not an isolated system—it is connected to the power plant that generates its electricity. The Second Law says that the increase in the orderliness of the water's atoms must be balanced by an increase in disorder in the atmosphere around the generating plant, an increase that is sure to result from waste heat.

Living systems are the most highly ordered form of matter we know. Staggering numbers of atoms must fit together in a precisely dictated way to make even the simplest cell. Creationists sometimes argue that evolution theories violate the Second Law because they assume that life appeared spontaneously. Nothing so ordered, they argue, could arise from disorder, because of the Second Law. Just as was the case with the ice cube, however, you have to consider the energy and randomness of the *entire system in which life arose*. That system includes not just the earth, but the earth's energy source—the sun—as well. The relative increase in order seen in living systems on the surface of our planet is more than balanced by the disorder created by the nuclear furnace that supplies the sun with energy; and the total entropy of the earth-sun system increases all the time.

The Second Law tells us that in nature, as in life, you have to pay for what you get. There is no free lunch!

FRONTIERS—NEW ENERGY SOURCES

Supplies of high-grade fuel, such as coal, natural gas, oil, and uranium, are limited and nonrenewable. A major research effort, therefore, involves the development of alternative energy sources. Power plants that rely on "renewable" energy—winds, tides, or solar radiation, for example—have been designed and built. At present only a small fraction of our energy needs are met by these technologies, partly because other power plants are well established, and partly because the alternative energy sources are expensive to implement.

The ultimate energy-producing technology is nuclear fusion, which imitates the Sun. A fusion reactor produces energy by converting hydrogen atoms into helium. In the process some mass is converted to heat energy, which can make steam and drive generators. The principal technological difficulty is that it takes tremendous pressures and temperatures (like those in the sun) to start the fusion reaction. Experimental fusion devices so far take more energy to operate than is produced by the nuclear reactions. With more research and improved reactor designs, however, fusion energy may contribute significantly to energy needs of the twenty-first century.

CHAPTER 3

ELECTRICITY AND MAGNETISM

6 A.M. Another weekday. The clock-radio blares, but you lie in bed a few minutes longer, listening to the news and weather and gathering energy to face the day. Turn on the light, start the coffee, wake the children, shower and dress. Grab the orange juice from the refrigerator—a note on the door reminds you that the kids have basketball practice after school. Eat a piece of toast, maybe some cereal. Brush your teeth, feed the cat, turn on the answering machine, and out the door to work. By 7 A.M. you're on your way to another busy day.

Gravity is not the only natural force you experience daily, nor even the strongest. In just one hour you've had dozens of run-ins with electricity and magnetism. A magnet clings to your refrigerator door, easily overcoming the gravitational force that the entire earth, pulling down, exerts on it. Static cling holds your clothes together, and you have to exert a force to pull them apart. These effects are not caused by gravity.

Electricity and magnetism are familiar forces and they appear everywhere in nature. Four laws of nature, Maxwell's equations, summarize everything we know about the phenomena of electricity and magnetism. The most important statement made in these equations is:

**Electricity and magnetism are two aspects
of the same force.**

Lightning, static cling, friction, TV transmission, and the little magnets you use to hold notes on your refrigerator are all siblings.

MAXWELL'S EQUATIONS

Once Isaac Newton demonstrated the power of the scientific method in mechanics, it was natural that the method would be applied to other areas. Eighteenth-century researchers with now familiar names like Alessandro Volta and André Marie Ampère studied electrical and magnetic phenomena as curiosities of the laboratory. They constructed batteries, examined the effects of electric sparks, passed current through various materials, and performed hundreds of other experiments. Driven by a desire to understand fascinating natural phenomena, these researchers never dreamed that electricity might someday transform society. Today we would say they were doing basic research.

The results of their experiments were summarized in laws, and these laws were brought together by the Scottish physicist James Clerk Maxwell in 1861. Maxwell's four equations, which in their abbreviated mathematical form have become a popular adornment of physics department sweatshirts, play the same role in electromagnetism that Newton's laws do for motion and gravity: they summarize everything there is to know on the subject.

Electrical Charge and Coulomb's Law

When you run a stack of papers through a photocopy machine, individual sheets may stick together. The force that holds the sheets together is said to be the electrical force, and objects that respond to the electrical force are said to have an electrical charge. Some simple experiments show that there are, in fact, two kinds of electrical charge. If you run two plastic combs through your hair, the force between them will be repulsive: they will be pushed apart. If you take one of those combs and bring it near a piece of glass that has been rubbed with fur, however, the objects experience an attractive force: they will be pulled together. There are two kinds of electrical forces, so it is reasonable to suppose they are generated by two kinds of electrical charge, which for historical reasons are called positive and negative.

Electrical charge is carried by subatomic particles—the building blocks of atoms. The atom's massive central nucleus is positively charged, while lighter, negatively charged electrons orbit the nucleus. An object is electrically charged if its atoms possess either an excess of electrons (in which case it has a negative charge) or a deficit of electrons (in which case it has a positive charge). In most situations the ponderous nuclei of atoms move very slowly, while electrons move easily. Thus, large objects usually acquire an electric charge by having electrons removed or added to their bulk. When you comb your hair, for example, electrons are stripped from your hair and pulled into the comb. As a result, the comb acquires a negative charge. This is why it will pick up bits of dust and paper (try this experiment yourself next time you comb your hair on a dry day). It also explains why the comb will then attract your hair and why the individual strands of hair repel each other—why they can "stand up." Vigorous combing can even make your hair stand on end as the deficit of electrons (and consequent positive charge) increases.

The French physicist Charles Coulomb (1736–1806) first wrote down the law that describes forces between electric charges:

Like charges repel each other; unlike charges attract.

and

Between any two charged objects is a force proportional to the size of the two charges, divided by the square of the distance between them.

This law says that if two objects have an excess of electrons (and therefore have negative charges), they will repel each other, but if one has an excess and one a deficit, they will attract. It also says that the form of the equation that describes the electrical force is strikingly similar to the one that describes gravity.

Coulomb's Law describes the force between electrical charges that do not move—what is called static electricity. Electrostatic forces dominate the world as we know it. Plus attracts minus in chemical bonds, and thus holds materials together. Every object you see is made from atoms, themselves collections of negative electrons attracted to positive nuclei. Just as the gravitational force keeps the earth and planets in orbit around the sun, electro-static attraction keeps negative electrons in orbit around the pos-itive nucleus of an atom.

The repulsion of electrons by electrons, on the other hand, keeps one object from passing through another. You can't put your hand through this book, for example, because electrons in atoms in your hand are repelled by electrons in atoms in the book. You don't fall through the floor, because electrons in your shoes repel electrons in the floor. Everytime you touch or feel something, you are making use of the electrostatic force.

Photocopies are products of electrostatic forces at work. A pol-ished plate of silicon metal can hold an electrical charge for ex-tended periods of time. When exposed to light, however, the charge leaks off. The key to xerography is to project a pattern of light and dark (such as a printed page) onto the charged plate. A similar pattern, consisting of charged and uncharged regions, is then created on the plate as charge leaves the lighted areas. Electrostatic forces cause a special black plastic powder to cling only to the charged areas of the silicon plate. The powder is transferred to paper, then melted in place, producing a copy of the original document.

Magnetism

Human beings have known about magnets and magnetism for thousands of years. Naturally occurring magnets, called lode-stones, were scientific curiosities in the ancient world, and slivers of lodestone that lined up in a north-south direction were the first

compasses. That a magnetic force exists can be verified by anyone who uses magnets to hang notes and miscellany on the refrigerator.

There is one common feature of all known magnets: each magnet, whether it be the size of an atom, a compass needle, or the
planet Earth itself, has two poles. Each pole is usually labeled
north or south, depending on which end of the earth they would
point to if they were allowed to act as a compass. Magnetic poles
have properties reminiscent of electric charge. Poles with the
same character always repel, while opposite poles always attract
(north attracts south, but repels north).

There is, however, an important difference between electrical
charges and magnetic poles—a difference enshrined in Maxwell's
second equation. No matter how hard you try, the law says, you
can never create an isolated magnetic pole. Unlike electrical
charges (which can exist as independent positive or negative particles), magnetic poles always come in pairs. If you cut a 2-inch-
long bar magnet in half, you don't get one north end and one
south end. You get two 1-inch-long bar magnets, each with its
own north and south pole. Cut those pieces in half and you just
get more magnets. Even the individual atoms are tiny "dipole"
(two-pole) magnets. Thus, Maxwell's second equation states:

There are no isolated magnetic poles.

Maxwell's second equation says nothing about how magnetic
fields come to be. Static electricity and magnetism seem to be
very different things, and there is no obvious connection between
a photocopy machine and refrigerator decorations. The nature of
magnetism, and the connection between it and electricity, is the
subject addressed in Maxwell's third and fourth equations.

Two Sides of the Same Coin

The relationship between electricity and magnetism can be stated
succinctly: every time an electric charge moves, a magnetic field

is created; and every time a magnetic field varies, an electric field is created. Electricity and magnetism are two inseparable aspects of one phenomenon: you cannot have one without the other.

If we could see or feel electric and magnetic fields, their close ties would be obvious because we'd always see them together. But in day-to-day life we are not usually aware of electrical effects when we use magnets, nor do we sense magnetic fields when we use electricity. We have to use instruments to tell us about the connection between the two.

The story of the discovery of this connection is a curious one. The Danish physicist Hans Oersted (1777–1851) was giving a physics lecture when he noticed that flipping a switch to start the flow of an electric current caused a nearby compass needle to twitch. Further experiments convinced him that a magnetic field is present whenever electrical charge flows through a wire. This finding (usually expressed in a suitable mathematical form) is Maxwell's third equation:

Moving electric charges create magnetic fields.

One common application of this law of nature is a device called the electromagnet. The simplest electromagnet is a loop of wire carrying an electric current. Because the current produces a magnetic field, the loop acts as a magnet. Unlike the permanent magnets that you use to hold things on your refrigerator, however, an electromagnet can be turned off and on by opening and closing the switch that controls the current.

A single loop produces a magnetic field with a north and a south pole. In fact, you can think of a current-carrying loop as equivalent to a small bar magnet with its north pole coinciding with the north end of the magnetic field created by the current. The only difference is that the polarity of the loop's field can be reversed by reversing the direction of the current. Electromagnets are found in many devices and machines, from ordinary doorbells to the large magnets that lift cars around in auto junkyards.

The other side of the electric/magnetic coin concerns the ability of magnetic fields to produce electrical forces. If the magnetic

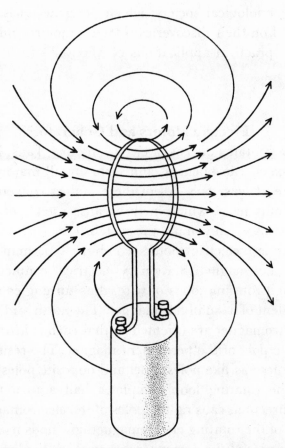

Electric current passing through a loop of wire creates a simple electromagnet, an essential component of every electric motor.

field in the region of a loop of wire is changed (by moving a magnet near the wire, for example), electrons will flow in the wire, even though there seems to be nothing in the wire to make them accelerate. This phenomenon, called electromagnetic induction, is described in the last of the Maxwell equations:

Magnetic effects can accelerate electrical charges.

Physicists Oersted, Henry, Faraday, and Maxwell did not know that their work would someday lead to large-scale generation and use of electricity. They could not have foreseen our twentieth-century technological society, which nevertheless is almost entirely based on their discoveries. Electric motors and generators are simply practical applications of Maxwell's third and fourth equations.

Electric Motors and Generators

Your home contains dozens of electric motors. Fans, hair dryers, razors, mixers, can openers, and virtually all major appliances incorporate at least one. All of these motors convert electricity into magnetic fields, which in turn cause useful rotary motion. What happens when you flip the switch?

The simplest electric motors combine a permanent magnet with an electromagnet. Stationary electrical contacts pass current through rotating loops of wire, thus turning each loop into the equivalent of a small bar magnet. The north and south poles of the electromagnet are oriented so that each is attracted to the appropriate pole of the permanent magnet. The result: the loop starts to rotate as like poles repel and opposite poles attract. As soon as the rotating loop completes half a turn the current switches direction, causing the poles of the electromagnet to flip. Each pole of the rotating electromagnet now finds itself attracted to the next pole of the permanent magnet, so the loop continues to rotate.

Most motors are more complex than the simple one described above. Typically, a motor incorporates multiple sets of permanent magnets or several synchronized electromagnets. Different arrangements of the basic components lead to motors that turn with a constant speed or a high torque or by small steps. In every case, however, the basic principle is the same: electricity is converted into magnetic fields.

Electrical generators are the exact opposite of electric motors: they convert rotary motion into electrical energy. The basic generator, first put to practical use by Thomas Edison, is little more than a loop of wire spinning in a magnetic field. Because of the rotation, the field seen by the loop is constantly changing, so a current flows in the wire, first one way and then the other. This

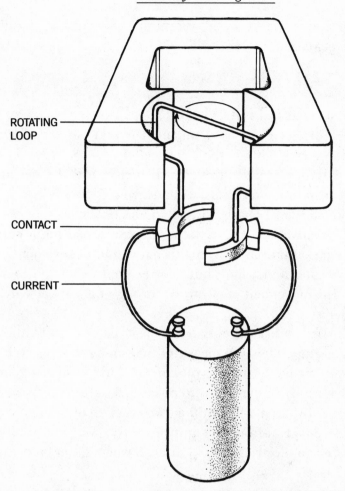

ROTATING LOOP

CONTACT

CURRENT

The simplest electric motor incorporates a permanent magnet and an electromagnet. Magnetic forces drive the rotary motion.

alternating (AC) current comes out of the generator on wires and can be used to run electrical circuits. Almost all electricity used in the United States is produced in this way.

Anything that can turn an axle can power a generator. Flowing water, pressurized steam, wind, or a gasoline engine can drive a rotating turbine that houses coils of copper wire. In a large generating plant, powerful electromagnets surround the wire loops. As copper wire cuts through the magnetic field lines, electrons are pushed back and forth—6o times per second in the United

States—to produce the 60-cycle alternating current that lights cities and runs air conditioners.

ELECTRICAL CIRCUITS

The aspect of electricity and magnetism that we encounter most often in everyday life is the electrical circuit. A circuit is a continuous path of material through which electrical charge can flow. The most common circuits are made from copper wires.

Any flow of electrical charges is called an electric current. Although both positive and negative charges can constitute a current, in everyday situations we give this name to the movement of electrons. A toaster and a light bulb, for example, both get their energy from the movement of electrons through the copper wires of your home.

Electric current is usually discussed in terms of a unit called the ampere, or amp. The amp measures how many charges go by a particular point in the wire (you can imagine a microscopic traffic counter sitting in the wire and pushing a button every time an electron goes by, then adding things up every second). Typical household currents run from one amp (in a 100-watt light bulb) to 50 amps (in an electric stove with all burners going and the oven on full blast).

Every circuit must also have a source—a device that supplies the energy to push electrons through the wires—which can be either a battery or a generator. In a battery, stored chemical energy is expended to provide the kinetic energy needed for electrons to move through the circuit. Current from a battery always flows in one direction, and is called DC (direct current). Current from a generator, on the other hand, alternates in direction and is called AC (alternating current).

The "pressure" with which electrons are pushed through a wire—the electrical potential—is measured in a unit called the volt. The higher the potential—the more volts—the more electrons can be pushed through a given wire. Some typical voltages encountered in everyday situations: flashlight batteries—1.5 V;

car batteries—12 V; household current—115 V; high voltage transmission lines—500,000 V.

ELECTROMAGNETIC RADIATION

The Nature of Light

Since all of the four experimental laws we have discussed so far were discovered by other people, you may be wondering why they are called Maxwell's equations. There are three reasons: (1) he was the first to see that the equations formed a coherent system; (2) he added a small piece to the Third Law (he proved that there was a kind of electrical current that no one had thought about up to that point); and (3) most important, he realized that the four equations predicted the existence of a new kind of energy wave— one that we now call electromagnetic radiation.

The third and fourth equations show that every field, magnetic or electric, induces a corresponding electric or magnetic field. Back and forth, *ad infinitum,* the fields create and modify each other. This sort of eternal oscillation, Maxwell realized, creates a wave that moves through space. Like ripples from a pebble thrown in a pond, these energy waves radiate out from their source.

All waves can be described by three closely linked characteristics: speed, wavelength, and frequency. Each wave consists of a series of crests and troughs. Wavelength is the distance between adjacent crests, speed is measured by the movement of the crests, and frequency is a measure of how many crests pass a given point in a second. The most common unit for measuring frequency is the hertz (named after the German radio pioneer Heinrich Hertz [1857–94]). One hertz (1Hz) corresponds to one crest going by a point each second. Look at the plate on any appliance in your home—it will say 60Hz, another reminder that household electrical current changes direction 60 times each second.

When Maxwell saw that his equations predicted the existence of waves, the first thing he did was calculate how fast those waves would move. He found that the speed of the mysterious waves depended on things like the force that one electrical charge or magnet exerts on another. These numbers had been measured in laboratories, so Maxwell was able to predict the velocity with high accuracy. His result: the waves move at 186,000 miles per second. This, of course, is what he knew, and we know, as the speed of light. Light itself is the mysterious electromagnetic wave.

The speed of light is so important that physicists denote it by a special letter— "c." It is the only speed that is actually built in to the laws of nature. It figures prominently in many fundamental theories, like the theory of relativity and its famous equation, $E = mc^2$. It also denotes the speed of other types of radiation like X rays and radio waves.

According to Maxwell's calculation, his waves were actually composed of electrical and magnetic fields alternately creating each other as they move through space. The frequency of an electromagnetic wave is simply the frequency of the oscillating field that caused it. If you wave a charged comb in the air once a second, for example, you create an electromagnetic wave with a frequency of 1 Hz and a wavelength of 186,000 miles. Atoms can vibrate a trillion times a second, giving waves about a hundredth of an inch in length. It takes a lot more energy to wiggle an electron trillions of times per second than just once, so higher-

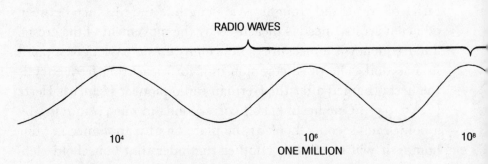

RADIO WAVES

10^4 10^6 10^8

ONE MILLION

Radio waves, microwaves, visible light, and X rays are all parts of the electromagnetic spectrum. Electromagnetic waves, which surround us all the time, are produced any time an electric charge accelerates.

frequency waves are also higher-energy waves. The important point, however, is that Maxwell's equations predict that electromagnetic waves should exist at all frequencies and all wavelengths, not just for the narrow band we call visible light.

The Electromagnetic Spectrum

Think about waves on the ocean. They range from mini-ripples to swells a few yards long to tidal effects that span the oceans. These ocean waves are all intrinsically the same, differing only in the size, frequency, and energy contained in the moving water. If you travel in an ocean liner you only notice a narrow range of these waves—the ones that make the ship rise and fall. Other waves are all around, but you don't sense them.

The same thing is true of electromagnetic radiation. Our eyes, like an ocean liner, sense a very narrow range of wavelengths— those around a few thousand atom diameters (about a ten thousandth of an inch)—but longer and shorter waves are all around us. The complete set of these waves is called the electromagnetic spectrum. All types of waves in this spectrum travel at 186,000 miles per second, and all are produced by moving electromagnetic fields.

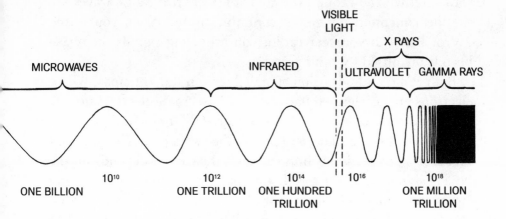

With his discovery Maxwell not only solved the mystery of the nature of light, but pointed to extraordinary practical consequences. As soon as he realized that visible light is only a narrow band of electromagnetic radiation, he postulated the existence of other waves of both longer and shorter wavelengths. These other waves included what we now call radio, microwave, infrared, ultraviolet, X rays, and gamma rays.

There is no theoretical limit to the wavelength of electromagnetic radiation; frequencies from zero to infinity are possible. In practice, however, we can only detect a limited range of waves, from radio waves a few thousand miles long to gamma rays with wavelengths smaller than atomic nucleii. Scientists and engineers have divided the spectrum into several regions, somewhat arbitrarily, based on how the radiation is produced and how it is detected.

RADIO WAVES

Radio waves encompass all electromagnetic radiation with wavelengths of a few yards to thousands of miles, the longest waves that we can easily produce and detect. Radio waves are very useful because they travel through air without being absorbed, they are easily generated and detected, and the longer wavelengths bend around the curve of the earth. Radio waves are the ideal medium for global communication. When you watch TV or listen to your car radio, you are using signals that have been transmitted by radio waves.

Both radio and television signals begin in tall antennas, where electrons are accelerated back and forth to create electromagnetic waves. All stations have a basic "carrier" frequency—the frequency of the wave that you read on the radio dial. The way that music or conversation is impressed on the carrier depends on the type of signal being sent. FM stations vary the frequency slightly (frequency modulation) while AM stations vary the signal strength (amplitude modulation). The difference between AM and FM is analogous to sending signals with a flashlight. If you send a signal by alternately dimming and brightening the flashlight, you are acting like an AM station. If, on the other hand, you send a signal by changing the color of the emitted light, you are like an FM station.

AM radio waves are about 1,000 feet in wavelength—long enough to bend around the curve of the earth. A strong station can be heard for hundreds of miles, especially at night when interference from other electromagnetic radiation is minimal. FM stations use radio waves only a few feet in wavelength. These waves do not bend around the earth, so FM stations must rely on line-of-sight transmission. This is why your favorite FM stations fade out when you drive more than about 50 miles from town. TV stations also usually transmit over the shorter wavelengths in the radio spectrum.

The radio part of the electromagnetic spectrum is wide, but it can only accommodate a finite number of separate channels. In addition to thousands of radio and television stations, there are hundreds of thousands of marine, aviation, amateur, and public safety broadcasters. The vast number of radio transmitters now in use would hopelessly clutter the airways, leading to electromagnetic chaos, without strict international controls over the allocation of broadcast frequencies, licensing, and operation of all radio stations. One of the principal responsibilities of the International Telecommunications Union and regional groups like the U.S. Federal Communications Commission is to allocate the long-wavelength end of the electromagnetic spectrum so that no two stations have frequencies that overlap.

MICROWAVES

Microwaves are electromagnetic waves about a tenth of an inch to a foot long. Longer microwaves, which have many features in common with radio waves, pass freely through air and can carry information. Unlike radio transmission, however, microwaves can be focused into a beam and therefore are often sent in highly directional signals, relayed with security from one cluster of hornlike antennas to another across the countryside. Furthermore, microwaves can be fine-tuned to yield a hundred times more useful frequencies than radio.

Line-of-sight transmission is essential for microwaves, so many microwave transmitters are prominently situated on tall towers or hilltops. More recently, microwaves have been used to communicate between the surface of the earth and satellites, which then beam the signal back to a different point on the earth. Over half the long-distance phone calls made in the United States are now routed through satellites via microwaves, as is satellite television. The TV dishes you see in rural backyards and hotel parking lots are all designed and carefully positioned to receive microwave signals sent down from satellites in fixed orbit. Commercial cellular phones operate in the same way, providing a link between a central transmitter and the mobile phone. In order to avoid cluttering the available channels, electronic systems break something like a large city into small units ("cells"), each with its own channel, and pass you along from one cell to the next, using whatever channel is available in each new cell.

Since World War II, microwaves have played a vital role in aircraft tracking. Radar employs directional microwave pulses, which reflect off solid objects in the air. The most sophisticated modern radar can pinpoint the location of a housefly at a distance of more than a mile.

Your microwave oven makes a very different use of electromagnetic radiation. The heart of the oven is a magnetron, a vacuum tube similar to your TV tube. A beam of electrons in the magnetron oscillates about two and a half trillion times a second to produce microwaves about five thousandths of an inch in wavelength. In your food, this particular radiation is absorbed by water molecules (clusters of two hydrogen atoms and one oxygen), which are then set into violent vibrations as the energy of the radiation is converted into molecular energy of motion. This molecular motion makes the food hot.

INFRARED RADIATION

The infrared portion of the electromagnetic spectrum extends from wavelengths of about a hundred thousandth of an inch to a tenth of an inch. The long wavelength end overlaps with micro-waves, while the short wavelength end stops at visible red light. Every warm object gives off infrared radiation. In the classic cow-boy movie, for example, the scout who holds his hands toward the remains of a campfire and announces that the bad guys are only an hour away is sensing infrared radiation emitted by the still cooling embers.

The infrared radiation we most commonly experience origi-nates from vibrations of molecules. When you sit in front of a fire the molecules of burning wood vibrate wildly, releasing heat ra-diation. That energy travels at the speed of light and is absorbed by your skin, setting your own molecules into vibration and trig-gering nerve impulses—you feel the heat.

Infrared radiation is absorbed in the atmosphere, so it is not very useful for communications. Even though our eyes can't see it, all objects absorb and emit infrared radiation. Each type of material has its own distinctive infrared "color," so many noctur-nal animals have developed infrared vision. Special infrared cam-eras in orbit around the Earth take advantage of the same phe-nomenon, as do the infrared "night" binoculars that allow hunters to see in the dark.

VISIBLE LIGHT

Visible light is the narrowest, but the most obvious, of the spec-tral regions. Most human eyes can detect waves between about 16 and 32 millionths of an inch long—roughly the distance across 5,000 atoms. Light is further divided into a spectrum of colors: red, orange, yellow, green, blue, indigo, and violet—liter-ally the colors of the rainbow. Of these colors, violet has the short-est wavelength (and therefore the highest frequency and energy), while red has the longest wavelength.

The importance of light to us sometimes makes it difficult to keep in mind its relative insignificance in the grand sweep of the electromagnetic spectrum.

ULTRAVIOLET RADIATION

Ultraviolet light starts at wavelengths just shorter than visible violet. This so-called "black light" is used in a wide range of applications. Precanceled postage stamps are tagged with fluorescent ink, theatrical productions incorporate colorful fluorescent paints, and amusement parks employ fluorescent hand stamps so visitors can come and go.

Shorter wavelength ultraviolet radiation has enough energy to disrupt and kill cells. Electromagnetic waves less than about a millionth of an inch in length are readily absorbed by living things and deposit sufficient energy to split apart molecules. For this reason, these wavelengths are routinely used in hospitals to sterilize equipment.

Ultraviolet radiation is absorbed in the atmosphere, particularly by ozone gas. The energetic radiation that leaks through this shield causes sunburn or even cancer on exposed skin. With UV radiation we being to enter the dangerous region of the electromagnetic spectrum.

X RAYS

Electromagnetic radiation with wavelengths about the size of an atom (a ten millionth of an inch) are called X rays. Their accidental discovery in 1895 revolutionized diagnostic medicine, since it gave physicians a chance to examine the interior of the human body without surgery.

X-ray machines at your doctor's or dentist's office are usually bulky metal things of odd dimensions, painted some depressing shade of green or gray. The workings are well concealed, but always contain two basic components that are housed in a high vacuum. At one end is a thin wire filament, similar to the one found in an ordinary incandescent light bulb. When heated to thousands of degrees, the filament emits a steady torrent of electrons, pulled out by strong electrical forces. The electrons are then accelerated toward a positively charged metal plate. They smash into the metal, and their deceleration unleashes a flood of energetic electromagnetic radiation—X rays.

The fact that X rays can go through solid matter makes them useful not only in medicine, but in the study of materials as well.

By studying how this sort of radiation interacts with a crystal, for example, scientists can deduce how the atoms inside the crystal are arranged.

GAMMA RAYS

Gamma rays, the most energetic electromagnetic radiation that we can measure, are produced in stars (cosmic rays) and during some radioactive decay. They have wavelengths much smaller than individual atoms and are therefore capable of passing through most solids. With the exception of a few medical testing procedures that use gamma-ray-emitting radioactive tracers, your life will probably be unaffected by these most energetic electromagnetic waves.

FRONTIERS

The Search for Magnetic Monopoles

The fact that magnetic poles always seem to come in pairs, never alone, has always seemed strange to physicists. In every other way, electricity and magnetism are mirror images of each other; only here do they differ. Consequently, physicists, trying to find what they believe to be a missing piece of nature's puzzle, have carried out many searches for "magnetic monopoles"—particles that carry an isolated magnetic "charge."

To date, all of these searches have been unsuccessful except for one event recorded in an apparatus at Stanford University in 1983. This event, as yet unexplained but also unduplicated, is in the limbo scientists reserve for results they can't prove wrong, but aren't ready to accept without confirmation.

If magnetic monopoles are eventually found, there will be no changes in the general features of electricity and magnetism as we experience them in everyday life. The main difference will be at the level of elementary particles.

CHAPTER 4

THE ATOM

What do the following things have in common?

an elephant
pantyhose
the Empire State Building
sand
your left ear
the Pacific Ocean
air
tofu
Jupiter
beer
this book

The answer is simple:

Everything is made of atoms.

Every tangible thing—the book you read, the food you eat, the air you breathe—is made of atoms. Atoms are the building blocks of matter. The atoms, in turn, are made largely from three types of smaller particles: protons and neutrons in the atomic nucleus, and electrons that orbit the nucleus. All of the amazing diversity of atoms—chemical elements as different as hydrogen, copper, sulfur, and uranium—results from different combinations of these three subatomic particles.

HOW DO WE KNOW THEY'RE THERE?

Atoms are a physical reality, but not one that you can verify just by looking around you. Atoms are so small that a million atoms placed end to end are no longer than a period on this page. The head of an ordinary pin contains more than 1,000,000,000,000,000,000,000 atoms. But although each atom has a minuscule mass, Nature has more than made up for the insignificance of each atom by producing a vast number of them.

The original idea of the atom is usually associated with the Greek philosopher Democritus, who lived sometime in the fifth century B.C. His argument went something like this: imagine that you have a very sharp knife and a piece of cheese. Cut a piece off the cheese, then cut that piece, then cut the resulting smaller piece, and so on and on. Two things might happen: you will either come to a smallest piece of cheese—the cheese atom—or you won't. Either possibility is reasonable. After all, you can build a house with individual bricks or from poured concrete. At a distance you see the house, but you can't tell how it is built. On philosophical grounds, Democritus argued that smallest bits must exist, and he gave them the name "atom"—that which cannot be divided. It wasn't until the early nineteenth century that the English chemist John Dalton (1766–1844) put forward our modern notion of the atom. Dalton was driven to believe in atoms by the results of laboratory experiments. Researchers discovered that most substances they encountered could be broken down in one way or another—by burning, by immersion in acid, or some other procedure. Occasionally, however, they would run across something that could not be broken down at all. Dalton called these substances, including oxygen, gold, sulfur, and iron, "elements."

Many common chemicals consist of precise ratios of elements. Water, whether taken from Arctic ice or tropical rain or distilled from living things, always has an exact ratio of hydrogen to oxygen of 1:8 by weight. Dalton guessed that each chemical element is represented by its own atom, and these atoms combine in simple ways. Water, for example, is made from two atoms of hydrogen and one of oxygen.

Throughout the nineteenth and early twentieth century, a debate went on over whether atoms are physically real or merely a useful idea: is matter really made from atoms, or does it just act as if it were? Atoms are much too small to see, so it was something like arguing about whether a whitewashed house in the distance is made from bricks or concrete. Albert Einstein ended this debate in 1905 when he explained a phenomenon called Brownian motion. When a small particle such as a grain of pollen is suspended in a liquid and observed under a microscope, it is seen to move around in a random, erratic path. Einstein explained that the particle moves because of collisions with atoms. Figments of the imagination can't produce motion, so Einstein argued that atoms must be real. Today, using devices called scanning tunneling microscopes, we can actually take "photographs" of individual atoms, so this old question has been firmly laid to rest.

Now scientists often argue about whether tiny particles inside the atoms are really made from even smaller particles, called "quarks," or just act as if they were—a debate that mirrors the old argument about atoms.

ANATOMY OF THE ATOM

The atom's structure closely parallels that of the solar system. A massive central nucleus, analogous to the sun, is orbited by smaller electrons, something like a swarm of planets. The nucleus has a positive electrical charge, the electrons a negative charge, and the electrical attraction between the two holds the whole system together.

The Nucleus

The nucleus is an extraordinary thing. It contains 99.9 percent of an atom's mass, but occupies only a trillionth of its volume. Atomic nuclei are tightly packed clusters composed primarily of protons and neutrons. These two atomic building blocks have nearly the same weight, and each weighs about 1,860 times more than an electron. Don't be taken in by the fact that protons and neutrons are "massive" on the atomic scale—it still takes about 600,000,000,000,000,000,000,000,000 of them to balance a thimbleful of water.

Protons determine how an atom will behave. Each proton has a positive electrical charge of $+1$, so the number of protons in the nucleus dictates the electrical characteristics of an atom. Each chemical element is defined exclusively by its number of protons—the so-called "atomic number." Every gold atom has exactly 79 protons. Helium, carbon, oxygen, and iron are element names for atoms with exactly 2, 6, 8, and 26 protons respectively. The number of other particles is irrelevant for the purposes of assigning names.

All the naturally occurring elements, from number 1 (hydrogen) to 94 (plutonium), are found in the rock, water, or air of earth. Of these ninety-odd elements, about fifty form almost everything that you are likely to see or use in a lifetime. Elements beyond number 94 can be created in specially equipped physics laboratories, although these "heavy atoms" are highly unstable and do not survive long. Elements heavier than 94 have names that honor prominent people and places of twentieth-century physics as in berkelium (97), einsteinium (99), and fermium (100).

Neutrons weigh roughly the same as protons but, lacking an electric charge, they have little effect on the structure of the atom or on the way one atom interacts with another. They play an important role, however, in holding the nucleus together, and they are as important as the proton in giving the atom mass. In fact, scientists weigh atoms and subtract the known weight of all the protons to determine how many neutrons are present.

The number of neutrons in a nucleus can vary from zero, in most hydrogen atoms, to more than 140 in the heaviest atoms. In most familiar atoms, protons and neutrons are present in roughly equal numbers. The commonest kind of carbon atom, for example, has 6 protons and 6 neutrons in its nucleus, while oxygen usually has 8 of each. In heavier elements, like iron (26 protons usually coupled with 30 neutrons) or platinum (78 protons and 117 neutrons), the neutrons outnumber protons by a modest margin.

The nuclei of a given chemical element must all have the same number of protons, but may have different numbers of neutrons. The nucleus of carbon, for example, usually has 6 of each, but occasionally there is a nucleus with 6 protons and 8 neutrons. An atom with this nucleus is still carbon, since it has six protons, but it weighs more than ordinary carbon. Atoms with nuclei that contain different numbers of neutrons are called isotopes of the element. Scientists customarily denote an isotope by giving the combined total of protons and neutrons. Thus ordinary carbon is called carbon-12, and the other kind carbon-14.

Electrons

We have already met electrons, the mobile carriers of negative electric charge. Electrons are tiny things, weighing only about 1/186oth as much as a proton or neutron. If you imagine an atomic nucleus as a basketball-sized 5-pound weight, then electrons are something like gnats flying around several miles away while everything in between is a void. Atoms, which form all "solid" matter, are themselves almost entirely empty space.

In order to keep the atom as a whole electrically neutral, the number of electrons in orbit must equal the number of protons in the nucleus exactly. Occasionally, because of collisions or some other mechanism, a particular atom may lose or gain electrons. An atom in which the number of electrons is not the same as the number of protons (and which is not, therefore, electrically neutral) is called an ion. The atoms in the gas in an ordinary fluorescent light bulb, for example, are often ionized when electrons are torn off normal atoms in collisions.

What You Need to Know

We've tried to stay away from jargon in this book, but there really are a few terms that a scientifically literate American should know. Here's a brief summary of key words relating to atoms—

words that appear again and again in news stories about nuclear waste, fusion power, superconductivity, and advances in electronics.

ATOM: The basic building block of everything around you.

NUCLEUS: The heavy central part of every atom; the nucleus contains protons and neutrons.

PROTON: A positively charged particle in the nucleus; the number of protons distinguishes one element from another.

NEUTRON: An electrically neutral particle in the nucleus.

ELECTRON: An electrically negative particle that orbits the nucleus.

ELEMENT: A substance that can't be broken down by chemical means; an atom for which you know the exact number of protons — the element carbon always has 6 protons.

ISOTOPE: An atom for which you know both the exact number of neutrons and the number of protons; the isotope carbon-14 always has 6 protons plus 8 neutrons. Different isotopes of a given element have the same chemical properties.

ION: An atom that has gained or lost electrons, and hence has an electrical charge.

THE BOHR ATOM

Although negative electrons circle the positive nucleus much as planets circle the sun, electron orbits differ from those of planets in one significant way. A planet like Earth does not have to be any specific distance from the sun — if its orbit were 10 feet closer to the sun or 10,000 miles farther out, no laws of physics would be violated. Electrons in an atom, however, can orbit only in certain well-defined paths, and can never be found anywhere else. Each of these so-called "allowed" orbits corresponds to a different energy, so the energy of atomic electrons can only have certain exact, specified values.

Electrons move from one orbit to another by means of a "quantum leap," a phenomenon that is impossible to picture. The electron simply disappears from one orbit and reappears in another *without traversing the space in between.* It's as if you moved up and down a staircase by simply vanishing from one step and popping up on the next. You may think that sounds like nonsense, but when we look at things in the universe that are extremely small, extremely massive, or traveling extremely fast, they just don't behave in familiar ways. In fact, many of the phenomena we encounter inside the atom have no analog in our everyday experience.

Since each electron orbit has a different energy, an electron moving to a new orbit must either give up or take in energy when it makes the leap. If the new orbit is closer to the nucleus than the old, the electron emits energy in the form of electromagnetic radiation. This is how atoms emit visible light. The light energy is precisely equal to the difference in energy between the old and new orbits. Similarly, if an atom takes in energy (by absorbing light, for example, or by collision with another atom), that energy can be given to the electron, which will then move to a higher orbit. This picture, in which electrons shuttle back and forth between allowed orbits as they absorb and emit energy, is called the "Bohr atom," after the Danish physicist Niels Bohr (1885–1962), who first suggested it as a young postdoctoral fellow in 1912.

THE PERIODIC TABLE
OF CHEMICAL ELEMENTS

When chemists first investigated how materials were put together, they made an important discovery. It turned out that you could take anything—a piece of wood, for example, or a stone—and break it down into other things. Wood, when burned, yields carbon dioxide, water, and some mineral ash. Each of these sub-

PERIODIC TABLE OF THE ELEMENTS

1 H Hydrogen																	2 He Helium
3 Li Lithium	4 Be Beryllium											5 B Boron	6 C Carbon	7 N Nitrogen	8 O Oxygen	9 F Fluorine	10 Ne Neon
11 Na Sodium	12 Mg Magnesium											13 Al Aluminum	14 Si Silicon	15 P Phosphorus	16 S Sulfur	17 Cl Chlorine	18 Ar Argon
19 K Potassium	20 Ca Calcium	21 Sc Scandium	22 Ti Titanium	23 V Vanadium	24 Cr Chromium	25 Mn Manganese	26 Fe Iron	27 Co Cobalt	28 Ni Nickel	29 Cu Copper	30 Zn Zinc	31 Ga Gallium	32 Ge Germanium	33 As Arsenic	34 Se Selenium	35 Br Bromine	36 Kr Krypton
37 Rb Rubidium	38 Sr Strontium	39 Y Yttrium	40 Zr Zirconium	41 Nb Niobium	42 Mo Molybdenum	43 Tc Technetium	44 Ru Ruthenium	45 Rh Rhodium	46 Pd Palladium	47 Ag Silver	48 Cd Cadmium	49 In Indium	50 Sn Tin	51 Sb Antimony	52 Te Tellurium	53 I Iodine	54 Xe Xenon
55 Cs Cesium	56 Ba Barium	57 La Lanthanum	72 Hf Hafnium	73 Ta Tantalum	74 W Tungsten	75 Re Rhenium	76 Os Osmium	77 Ir Iridium	78 Pt Platinum	79 Au Gold	80 Hg Mercury	81 Tl Thallium	82 Pb Lead	83 Bi Bismuth	84 Po Polonium	85 At Astatine	86 Rn Radon
87 Fr Francium	88 Ra Radium	89 Ac Actinium	104 Unq (Unniliquadium)	105 Unp (Unnilpentium)	106 Unh (Unnilhexium)												

58 Ce Cerium	59 Pr Praseodymium	60 Nd Neodymium	61 Pm Promethium	62 Sm Samarium	63 Eu Europium	64 Gd Gadolinium	65 Tb Terbium	66 Dy Dysprosium	67 Ho Holmium	68 Er Erbium	69 Tm Thulium	70 Yb Ytterbium	71 Lu Lutetium
90 Th Thorium	91 Pa Protactinium	92 U Uranium	93 Np Neptunium	94 Pu Plutonium	95 Am Americium	96 Cm Curium	97 Bk Berkelium	98 Cf Californium	99 Es Einsteinium	100 Fm Fermium	101 Md Mendelevium	102 No Nobelium	103 Lr Lawrencium

The periodic table provides a separate box for each chemical element, with all the elements in one vertical column having similar properties. Each box shows the element's number (the number of protons in the nucleus) and the standard one- or two-letter abbreviation.

stances in turn can be broken down by the appropriate procedures into still other substances—carbon, hydrogen, and water,
for example. But try as they would, chemists could not find a way
to break things like carbon and oxygen down further. These substances, which seemed to make up the pieces of more complex
materials but which were themselves irreducible, scientists
termed "chemical elements." At the end of the eighteenth century chemists knew of about twenty-six elements, and today that
list has expanded to just over a hundred.

Every element has a name. "Hydrogen," for example, comes
from the French for "generator of water," a name that tells the
story of how the element was discovered. We customarily represent each element by a simple one- or two-letter abbreviation—
H for hydrogen, O for oxygen, Ca for calcium, and so on.

Chemists systematize all the elements in one simple table—the
periodic table. Every element has its own box, with atomic number increasing as you read from left to right. Boxes are arranged
so that elements in a vertical grouping have similar chemical behavior—that is, elements in the same column enter into similar
reactions and combine to form similar compounds.

The periodic table of the elements, a fixture in chemistry lecture halls, was first written down by the Russian scientist Dmitri
Mendeleyev (1834–1907) in 1869. He didn't understand why a
ranking of the elements in increasing order of weight had something to do with their chemical properties, but it seemed that it
did. In addition, when Mendeleyev first wrote down the table
there were two holes, corresponding to the elements we now call
germanium and scandium. When these elements were discovered
(in Germany and Sweden, as the names imply), it was considered
an important piece of evidence that the organization of Mendeleyev's table has a deep reason behind it.

Today we understand that elements can be grouped into rows
and columns because of the way electrons fit into Bohr orbits of
atoms. It turns out that electrons cannot be crowded too closely
together. Like two cars in a parking lot, two electrons cannot
share the same space. We call this the exclusion principle (because the presence of one electron excludes others).

Thus, there is space for only two electrons in the lowest Bohr orbit. The atoms of hydrogen (one electron) and helium (two electrons) respectively fill that orbit, so the inner electron shell becomes "closed," to use physicists' jargon. The next element, lithium, has three electrons, however, so the third electron must go into the next higher orbit. Thus lithium, like hydrogen, has a single electron in its outermost orbit, which explains why the two have similar chemical properties.

The second and third orbits have space for eight electrons each, so we can squeeze in seven electrons after lithium before we start on another shell, and thus find another atom with one electron in its outermost orbit. This element is sodium, with eleven electrons. Not surprisingly, you will discover hydrogen, lithium, and sodium in the same column of the periodic table, along with potassium, rubidium, cesium and francium—all of which have one electron in their outer Bohr orbits.

Using quantum mechanics, the subject of the next chapter, we can predict the number of spaces available for electrons in each atomic orbit. It is this calculation that ultimately justifies and explains the periodic table.

FRONTIERS

Photographing the Atom

One of the most striking areas of modern research is the rapid development of "microscopes" capable of producing photographs of individual atoms in a material. The best-developed of these instruments, the scanning tunneling microscope (or STM), works by measuring the electric current that flows between a tiny, precisely positioned point and the atoms on the surface of a material. The closer the point of the atom, the greater the current. A typical STM photograph shows the presence of individual atoms on a surface.

As scientists improve the resolution of these microscopes, and as they learn more and more about the surfaces of materials all around us—metals, plastics, paper, and skin, to name just a few—we can expect to see more spectacular pictures of atoms.

A scanning tunneling microscope reveals the locations of individual iodine atoms coated on a surface of platinum. The distance between atoms is only about a billionth of an inch. PHOTO COURTESY OF BRUCE SCHARDT, PURDUE UNIVERSITY.

Superheavy Atoms

Although natural elements occur only up to atomic number 94, artificial elements up to 109 have been created in the laboratory. These atoms, with their heavy collection of protons and neutrons, have unstable nuclei and break down quickly into groups of simpler atoms. Since the discovery of one atom of element 109 in a laboratory in Darmstadt, Germany, in 1982, the progress in the creation of new chemical elements seems to have stalled while a new round of experimental machines are being built. Theorists think that when we get to atomic numbers in the region of 120 to 130, we will find a new group of stable elements—elements to which they give the name "superheavy." If they are right, the periodic table of the twenty-first century may be a lot longer than the one we have now.

CHAPTER 5

THE WORLD
OF THE QUANTUM

The word "quantum" is familiar to lay people—splashed on cars and bandied about by Madison Avenue. Pundits and newscasters talk about "quantum leaps," mostly in a context that has nothing to do with physics. But the ideas of quantum mechanics are neither familiar nor obvious, and the term can instill fear even in the hearts of otherwise knowledgeable scientists.

Most scientists understand quantum mechanics in a vague, nonmathematical sort of way, but very few can do anything practical with it. That's because quantum mechanics can be devilishly hard in all its quantitative rigor. Yet, in spite of this aura of complexity, the two basic ideas behind quantum mechanics—what you need to know to be scientifically literate—are quite simple:

Everything—particles, energy, the rate of electron spin—comes in discrete units, and you can't measure anything without changing it.

Together, these two basic facts explain the operation of atoms, things inside atoms, and things inside things inside atoms.

THE WORLD OF THE VERY SMALL

Quantum mechanics is the branch of science devoted to the study of the behavior of atoms and their constituents. *Quantum* is the Latin word for "so much" or "bundle," and "mechanics" is the old term for the study of motion. Thus, quantum mechanics is the study of the motion of things that come in little bundles.

A particle like the electron must come in a "quantized" form. You can have one electron, or two, or three, but never 1.5 or 2.7. It's not so obvious that something we usually think of as continuous, like light, comes in this form as well. In fact, the quantum or bundle of light is called the "photon" (you may find it useful to remember the "photon torpedoes" of *Star Trek* fame). It is even less obvious that quantities like energy and how fast electrons spin come only in discrete bundles as well, but they do. In the quantum world, *everything* is quantized and can be increased or decreased only in discrete steps.

The behavior of quanta is puzzling at first. The obvious expectation is that when we look at things like electrons, we should find that they behave like microscopic billiard balls—that the world of the very small should behave in pretty much the same way as the ordinary world we experience every day. But an expectation is not the same as a commandment. We can *expect* the quantum world to be familiar to us, but if it turns out not to be, that doesn't mean nature is somehow weird or mystical. It just means that things are arranged in such a way that what is "normal" for us at the scale of billiard balls is not "normal" for the universe at the scale of the atom.

THE UNCERTAINTY PRINCIPLE

The strangeness of the quantum world is especially evident in the operation of the Uncertainty Principle, sometimes called the Heisenberg Uncertainty Principle after its discoverer, the German physicist Werner Heisenberg (1901–76). The easiest way to understand the Uncertainty Principle is to think about what it means to say that you "see" something. In order for you to see these words, for example, light from some source (the sun or a lamp) must strike the book and then travel to your eye. A complex chemical process in your retina converts the energy of the light into a signal that travels to your brain.

Think about the interaction of light with the book. When you look at the book you do not see it recoil when the light hits it, despite the fact that floods of protons must be bouncing off every second in order for you to see the page. This is the classic Newtonian way of thinking about measurement. It is assumed that the act of measurement (in this case, the act of bouncing light off the book) does not affect the object being measured in any way. Given the infinitesimal energy of the light compared to the energy required to move the book, this is certainly a reasonable way to look at things. After all, baseballs do not jitter around in the air because photographers are using flashbulbs, nor does the furniture in your living room jump every time you turn on the light.

But does this comfortable, reasonable, Newtonian viewpoint apply in the ultra-small world of the atom? Can you "see" an electron in the same way that you see this book?

If you think about this question for a moment, you will realize that there is a fundamental difference between "seeing" these two objects. You see a book by bouncing light off it, and the light has a negligible effect on the book. You "see" an electron, on the other hand, by bouncing another electron (or some other comparable bundle) off it. In this case, the thing being probed and the thing doing the probing are comparable in every way, and the interaction cannot leave the original electron unchanged. It's as if the only way you could see a billiard ball was to hit it with another billiard ball.

There is a useful analogy that will help you think about measurement in the quantum domain. Suppose there were a long, dark tunnel in a mountain and you wanted to know whether there was a car in the tunnel. Suppose further that you couldn't go into the tunnel yourself or shine a light down it—that the only way you could answer your question was to send another car down that tunnel and then listen for a crash. If you heard a crash, you could certainly say that that was another car in the tunnel. You couldn't say, however, that the car was the same after your "measurement" as it was before. The very act of measuring—in this case the collision of one car with the other—changes the original car. If you then sent another car down the tunnel to make a second measurement, you would no longer be measuring the original car, but the original car as it has been altered by the first measurement.

In the same way, the fact that to make a measurement on an electron requires the same sort of disruptive interaction means that the electron (or any other quantum particle) must be changed whenever it is measured. This simple fact is the basis for the uncertainty principle and, in the end, for many of the differences that exist between the familiar world and the world of the quantum.

The uncertainty principle is a statement that says, in effect, that the changes caused by the act of measurement make it impossible to know everything about a quantum particle with infinite precision. It says, for example, that you cannot know both the position (where something is) and velocity (how fast it's moving) exactly—the two pieces of data that are significant in describing any physical object.

The important thing about the uncertainty principle is that if you measure the position of a tiny particle with more and more precision, so that the error becomes smaller and smaller, the uncertainty in velocity must become greater to compensate. The more care you take to know one thing, the more poorly you know the other. The very act of measuring changes the thing you are measuring, so you must always be uncertain about something.

THE WAVE FUNCTION

The fact that one cannot measure a quantum system without changing it leads to an extremely important conclusion about the way that such systems must be described. Suppose that large objects, like airplanes, behaved the way electrons do. Suppose you knew that an airplane was flying somewhere in the Midwest and you wanted to predict where it would be a few hours later. Because of the uncertainty principle, you couldn't know both where the plane was and how fast it was going, and you'd have to make some compromise. You might, for example, locate the plane to within fifty miles and its velocity to within 100 miles per hour.

If you now ask where the plane will be in two hours, the only answer you can give is "It depends." If the plane is traveling 500 miles per hour, it will be a thousand miles away; if it's traveling 400 miles per hour, it will be only 800 miles away. And, since we don't know exactly where the plane started, there is an additional uncertainty about its final location.

One way to deal with this problem is to describe the final location of the plane in terms of probabilities: there's a 30 percent chance it will be in Pittsburgh, a 20 percent chance it will be in New York, and so on. You could, in fact, draw a graph that would show the probable location of the plane at any point east of the Mississippi. For historical reasons, a collection of probabilities like this is called the "wave function" of the plane.

We normally don't worry about wave function for airplanes, because in the everyday world the amount of change caused by a measurement is so small as to be negligible, so the uncertainties in the plane's position and speed are tiny. In the quantum world, however, *every* measurement causes appreciable change in the object being measured, and hence everything has to be described in terms of probabilities and wave functions. It was this unpredictable aspect of the quantum world that troubled Albert Einstein and caused him to remark that "God does not play dice with the universe." (His old friend Niels Bohr is supposed to have replied, "Albert, stop telling God what to do.")

WAVES OR PARTICLES?

We all run into trouble when we try to visualize a quantum object like an electron. Our inventory of mental images is limited to the sorts of things we can see in our familiar world, and, unfor-

tunately, electrons just don't fit anywhere on our mental file
cards. Nowhere is this problem of visualization more difficult
than in discussion of particles and waves in the quantum world.

In our normal world, energy can be transferred by particle or
by wave, as you can see by thinking about a bowling alley. Suppose there were one pin standing at the other end and you wanted
to knock it down. You would have to apply energy to the pin to
do this, of course, and might choose to expend some energy to get
a bowling ball moving, and then let the ball roll down the alley
and knock the pin over. This process involves a particle—the
bowling ball—carrying energy from you to the pin. Alternatively,
you could set up a line of pins and then knock the first one over.
It would knock over the second, which would knock over the
third, and so on like dominoes until the final pin fell. In this case,
the energy would be transmitted by the wave of falling pins, and
no single object travels from you to your target.

When scientists started to explore the subatomic world, they
naturally asked "Are electrons particles or are they waves?" After
all, an electron transfers energy, and if energy can be transferred
only by particles and waves, then the electron must be one or the
other.

Unfortunately, things aren't that simple. Experiments performed on electrons have found that in some situations they seem
to act as particles, in other situations as waves. Similarly, something we normally consider to be a wave—light, for example—
can appear to be a particle under certain circumstances. In the
early years of this century, this seemingly inexplicable behavior
was called "wave-particle duality" and was supposed to illustrate
the strangeness of the quantum world.

There is, however, nothing particularly mysterious about "duality." The behavior of electrons and light simply tells us that in
the quantum world, our familiar categories of "wave" and "particle" do not apply. The electron is not a wave, and it's not a
particle—it's something else entirely. Depending on the experiment we do, we can see wave-like or particle-like behavior. The
wave-particle problem lies not with nature, but with our own
minds.

Suppose you were a Martian who, for some reason or other, had been able to pick up radio broadcasts from Earth only in the French and German languages. You might very well come up with a theory that every language on Earth was either French or German. Suppose then that you came to Earth and landed in the middle of an American city. You hear English for the first time, and you note that some of the words are like French and some are like German. You would have no problem if you realized that there was a third type of language of which you had been previously ignorant, but you could easily tie yourself in philosophical knots if you didn't. You might even develop a theory of "French-German duality" to explain the new phenomenon.

In the same way, as long as we are willing to accept that things at the atomic level are not like things in our everyday world, no problem arises with the question of whether things are waves or particles. The correct answer to the wave-particle question is simply "none of the above."

Of course, this means that we cannot visualize what an electron is like—we can't draw a picture of it. For creatures as wedded to visual imagery as we are, this is deeply troubling. Physicists and non-physicists alike rebel and try to make mental images, whether they are "real" or not. The authors are no different, and, for the record, we imagine the electron as something like a tidal wave, located in one general area, like a particle, but with crests and troughs, like a wave.

The length of the "tidal wave" associated with different kinds of particles varies tremendously. That of an electron, for example, is smaller than that of an atom, while a photon of ordinary visible light is about three feet long. Viewed this way, both "waves" like light and "particles" like electrons have the same basic structure. The distinction in classical physics between wave and particle turns out to be a meaningless distinction in the quantum world.

THE ATOM—QUANTUM MECHANICS IN ACTION

The most important role quantum mechanics plays in science is explaining how the atom is put together. In the last chapter we described the peculiar property of electrons in the Bohr atom to adopt fixed orbits. These fixed orbits are a consequence of quantized electron energies. Electrons can only have certain precise energies, and any quantum leap between orbits must correspond exactly to the difference in orbital energies. Each quantum leap by one electron leads to the absorption or the production of one photon.

Electrons moving up and down in their orbits are analogous to a person moving up and down a staircase: it requires energy to climb, and energy is released upon descent. And, like a person on a staircase, an electron cannot be found between "steps"—in other words, it can only be found in allowed orbits.

Although it is tempting to think of electrons in orbits as particles, little lumps of matter, scientists often picture them in terms of their wave functions. The peak in the electron "wave," corresponding to the highest probability of finding the electron, is at the place where the electron would be located if we pictured it as a particle.

Lasers

From grocery lines to rock concerts, compact discs to the most advanced weaponry, lasers are changing our world by changing the ways we use light.

"Laser" is an acronym for Light Amplification by Stimulated Emission of Radiation, an imposing name for a remarkable device. Lasers work like this: you start off with a collection of atoms, each with an electron in a high-energy orbit. The chromium atoms in crystals of ruby serve this function in many red lasers. Photons with exactly the same energy as the excited electrons are focused on those special atoms. When one of these photons comes near an atom it "stimulates" the electron in the atom to jump down, emitting another photon in the process—one that is not only of the same wavelength as the original, but is precisely aligned, crest-to-crest, trough-to-trough. The two photons now pass through the material, stimulating other atoms until a flood of precisely aligned photons results. In this way, one photon "amplifies" itself.

The energy needed to get the atoms into an excited state in the first place, and to get them to go back to it after they have emitted a photon, can be added to the system in many ways. Typically, scientists "pump" a laser by subjecting the material to heat, to a beam of energetic electrons, or to a bright light, from something like a flashbulb or even another laser.

Two precisely aligned mirrors at each end of the laser material cause the photons to move back and forth millions of times. Engineers design laser mirrors to allow a small fraction (maybe 5 percent) of the photons to escape on each bounce, and these left-over photons form the laser beam.

FRONTIERS

Visualizing the Quantum

There is a group of physicists who just can't seem to leave the quantum alone. Like a dog worrying a bone, they keep coming back to central issues like wave-particle duality, introducing new and highly innovative experiments to shed light on this (and other) issues of "quantum weirdness." The press, often reports on these experiments because their results invariably present the kind of conflict between science and intuition we talked about earlier.

There are a large number of experiments being done in which the experimenters try to "trick" the quantum into revealing its true identity. For example, scientists have designed experiments that delay the decision of whether to test the wave- or particle-like aspects of a neutron until the neutron is actually in flight toward the apparatus. (For technical reasons, the electrically neutral neutron provides the best particles for these tests.) The results of such experiments are invariably in line with quantum mechanics—the "wave" experiment sees the neutron as a wave, while the "particle" experiment sees it as a particle. Nevertheless, the results run so counter to our intuition that they compel attention. It may be the only situation in science where we are more troubled when an experiment agrees with our theory than when it doesn't.

Quantum Detectors

Ronald Reagan often spiced his arms reduction rhetoric with the phrase "trust but verify." When pressed, Reagan revealed that verification would be accomplished, in part, by "technical means." Part of what he was really talking about were the most sensitive possible detectors, based on the principles of quantum mechanics.

The most sensitive detector responds to a single quantum—one photon, for instance. Among the most important of these tools is the SQUID, or Superconducting Quantum Interference Device. The SQUID stores a precise amount of electrical current in a tiny wire loop. Any slight change in magnetic field alters the current in the loop by one quantum unit, and that in turn triggers the detector. SQUIDs play a vital role in science, in medicine, and in detecting the small magnetic disturbances caused by a moving enemy submarine.

CHAPTER 6

CHEMICAL BONDING

Nothing beats Margee's Pennsylvania applesauce bread.

> 2 cups all-purpose flour
> ¾ cup sugar
> 1 teaspoon baking powder
> 1 teaspoon baking soda
> 1 teaspoon cinnamon
> ½ teaspoon nutmeg
> 1 teaspoon vanilla
> ½ cup shortening
> 1 cup applesauce
> 2 eggs

Combine all ingredients in large mixing bowl. Beat at medium speed until well blended. Grease a 9″ × 5″ loaf pan on the bottom only, and pour in mixture. Bake at 350° for 55 to 60 minutes, until a toothpick comes out clean. Loosen the edges with a spatula and remove from the pan. Cool before slicing.

Scientists like to cook up new recipes. Early in 1986 two scientists made history (and won the Nobel Prize) by mixing and grinding ordinary elements like copper, oxygen, and a couple of others in just the right proportions, and baking them at just the right temperature just long enough to produce a nondescript black wafer about the size of an aspirin tablet. That little black disk turned out to be the first of an entirely new kind of superconductor, a material with extraordinarily valuable electrical properties.

How is it possible that ingredients as different as flour, eggs, salt, and applesauce can combine to form a delicious loaf of bread? How can commonplace elements as different as copper and oxygen become a priceless black disk? The answer lies in the atoms—the building blocks of everything around us. Atoms combine in countless ways to form materials with every imaginable property. But in every case:

Atoms are bound together by electron glue.

The properties of all materials, be they sticky, magnetic, brittle, or green, arise from the infinite different ways atoms can be arranged and linked together. Two atoms form a chemical bond when their electrons rearrange themselves so that each atom feels an attractive force. The attraction of positive and negative electrical charges holds everything together.

THE ELECTRON GLUE

Chemistry is first and foremost the science of electrons and their interactions. That may seem an odd assertion, for chemistry is usually portrayed as the science of test tubes and bubbling beakers and odd mixtures of stuff that turns blue. But all those chemical reactions are the results of electrons shifting between atoms.

When two atoms approach each other, their outermost electrons come in close contact; the two negative electric charges repel each other. In most instances, as in the collision of gases in the atmosphere, the atoms just career off to another chance rendezvous. Occasionally, however, colliding atoms stick together by an exchange or sharing of electrons.

The first step in understanding the extraordinary variety of materials that surround us is to look at the way that two or more atoms can link together: the chemical bond. There are several different kinds of bonds, each producing dramatically different properties. The general rule that dictates what kind of linkage, if any, will be formed when two atoms come near each other is simple: the two-atom system will try to reach a state of lowest possible energy. Just as a ball rolls down a hill to get to a position of lowest gravitational potential energy, an atom's electrons rearrange themselves to reach a state of lowest electrical potential and kinetic energy.

Four types of chemical bonds hold almost all materials together. Each involves a different rearrangement of electrons between atoms. The distinctive materials that enrich our world can, in fact, be described in architectural terms. Various atomic building blocks—atoms—are joined in structures using four different kinds of glue—chemical bonds.

Elements in Combination

Elements can be combined with each other in countless ways to form chemical compounds that have properties and applications very different from their elemental raw materials. Nature and chemists have created millions of different chemicals. Some are very simple, like H_2O (water) with two hydrogen atoms for every oxygen, or $NaCl$ (table salt) with a one-to-one ratio of sodium and chlorine. Other compounds are exceedingly complex, combining a dozen or more different elements.

With fewer than one hundred elements to play with, it might seem that the chemist's job of making new chemical combinations would be quickly exhausted. Not so. A million chemists could each make a new sample every day for a million years and still not come close to running out of things to try. There are about 70,000,000 possible four-element combinations in one-to-one-to-one-to-one ratios, to say nothing of the countless variations in element ratios. For each possible composition there are hundreds of ways to mix and treat the chemicals, each resulting in a different arrangement of the same set of constituent atoms, and therefore in a material with different properties. Just as in

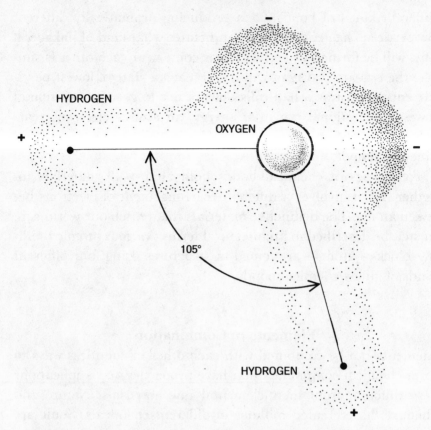

Each boomerang-shaped molecule of water (H_2O) contains a central atom of oxygen flanked by two smaller atoms of hydrogen.

cooking, the temperature and time of baking can make all the difference between a culinary masterpiece and an unmitigated disaster. The successful chemist, like a great chef, must apply skill and intuition in preparing new recipes.

Chemists make their living by trying to create new and useful mixtures of atoms, or by trying to manufacture established useful chemicals in new ways. This is not an idle pursuit. Almost every aspect of modern life—food and clothing, transportation and communications, sports and entertainment—depends on the discoveries of chemistry.

The Ionic Bond

An ionic bond forms when one atom in a pair gives up an electron and the other acquires it on permanent loan. In common table salt (sodium chloride), for example, the sodium gives up its outermost electron and the chlorine takes it. When this happens, each previously neutral atom becomes electrically charged—an ion (hence the name ionic bond). The sodium ion has a positive charge, the chlorine a negative one, and the two atoms are held together by the electrostatic attraction of opposite charges.

The formation of the ionic bond can best be thought of in terms of electron shells. The number of electrons that can be put into any orbit, or shell, is limited by the laws of quantum mechanics. From the point of view of the atom, the ideal state—the one with lowest energy—is the one in which every available space in an orbit is filled with an electron, and no electron gaps exist. Scientists call this "full up" situation a closed shell.

Atoms that normally have closed shells, like helium or neon, are extremely stable and do not react chemically with each other or anything else. Other elements do not share this complacence. Chlorine, for example, falls one electron short of having a filled shell, while sodium has one electron too many. When sodium and chlorine meet, an electron is exchanged and each atom ends up with a filled electron shell.

Hundreds of everyday materials are held together by ionic bonds. Table salt is the classic textbook example of ionic bonding, but compounds that combine silicon and oxygen, two of the earth's most abundant elements, are by far the most common ionic compounds. Most beach sand is a mineral called quartz, formed when silicon atoms, which yield four electrons, bond to pairs of oxygen atoms, each of which collects two electrons, leaving both silicon and oxygen with closed shells. The large electrostatic charges of $+4$ on every silicon and -2 on every oxygen results in a strong electrostatic force between adjacent atoms. That is why quartz, which can scratch steel, is such a tough and resistant mineral. Similar ionic silicon-oxygen bonds provide the strength and hardness of window glass, china and other ceramics, and most common rocks and minerals.

The Covalent Bond

A curious dilemma faces two carbon atoms as they approach each other. Each carbon has an outer shell that is exactly half full. Is the atom supposed to donate or accept electrons? In fact, neither atom does anything of the sort. The two carbons actually end up sharing their outer electrons—in effect creating a filled outer shell between both atoms. Electrons, constantly flipping back and forth between adjacent atoms, do not really belong to either carbon. This sharing produces a force that holds the two atoms together—a covalent bond. Although the covalent bond is most often seen holding carbon atoms together, many other elements, including silicon, sulfur, and nitrogen, participate in this sort of bonding as well.

The covalent bond is the basis of all life. It holds the tissues of your body together and keeps your DNA from falling apart. It also can be seen in plastics, nylon, diamonds, and superglue. Look around you. Carbon-carbon bonding is an essential part of everything you see, except for metals, water, glass, and ceramics. Carbon-carbon bonding is so important that tens of thousands of specialists, known as organic chemists, devote their research lives exclusively to the study of carbon compounds.

The main reason that the covalent bond is seen so often in nature is that covalent compounds seldom stop at a single pair of atoms. If we start with two carbon atoms sharing a pair of electrons, for example, other carbon atoms can start to share other electrons, forming ever longer and more complex structures. Our world is full of giant atomic structures with thousands of carbon atoms in long chains or intricate branches or interlocking ring patterns. There is literally no limit to the variety of potential organic compounds. In all of these substances, no matter how fancy, each pair of adjacent carbon atoms shares some electrons in a covalent bond.

The Metallic Bond

Atoms in metals generally have filled electron shells plus one or more electrons left over. Metallic sodium, a soft silvery metal, has one electron outside of a closed shell, magnesium has two, aluminum three, and so on. When metal atoms combine, some or all of these overflow electrons leave their homes and wander freely throughout the metal, leaving behind positively charged ions as they swim in a sea of negative charge. Each nucleus creates a local island of positive charge, and electrostatic forces hold the whole metal system together. You can think of the metallic bond as a system in which outer electrons are shared by all the atoms in the system, in contrast to the covalent bond, in which only neighboring atoms share a given electron.

Metallic bonds differ dramatically from ionic bonds. In an ionic crystal there are always two very different kinds of atoms, some with positive charge and some with negative charge. Each ion is surrounded by those of opposite charge. Obviously, a pure element (with only one kind of atom) can never have ionic bonds. In a metal, however, all atoms play exactly the same role as their neighbors. Every metal atom is surrounded by similar metal atoms. It's probably not surprising, therefore, that about three quarters of all known pure elements, including iron, aluminum, copper, and gold, form metallic bonds. Metallurgists often mix two or more of these elements to create metal alloys with distinctive properties. Brass combines copper and zinc, bronze forms from copper and tin, and steel alloys can incorporate a dozen metallic elements along with the essential iron and carbon components.

Chemical Bonds and the Real World

Chemical bonds in lots of materials don't exactly fit the neat definitions of ionic, covalent, or metallic. Electrons are always on the move, and it is not always possible to say exactly where a given particle spends its time. If outer electrons remain near one ion then the bonding is ionic; if they are shared equally between a pair of atoms then the bonding is covalent; if they are free to roam the crystal then the bonding is metallic. But in many substances electrons divide their time unevenly between two or more atoms. Such divided loyalties result in bonds with a mixed character.

The common mineral pyrite, known to miners as "fool's gold," owes its unusual combination of properties to iron-sulfur bonds that have a complex mixture of covalent, metallic, and ionic character. Fool's gold's shiny luster confused thousands of novice California prospectors during the 1849 gold rush, but the mineral breaks with rough, brittle edges, characteristic of materials with ionic or covalent bonding.

The van der Waals Bond

In thousands of compounds groups of atoms bond without ever giving up an electron, a situation very different from ionic, metallic, or covalent bonding. This type of cohesive force arises because when two atoms come near each other, their electron clouds distort as a result of the repulsive force between electrons. The electrical forces between two distorted atoms are (1) repulsion between the two nuclei, (2) attraction between the electrons of one atom and the nucleus of the other, and (3) repulsion between the electrons. Although it isn't obvious, it turns out that in this situation the attractive forces can win, in which case a weak bond forms between the atoms. We call this bond the van der Waals force, after Dutch physicist Johannes van der Waals (1837–1923). Many soft materials, from candle wax to talcum powder, owe their distinctive properties to weak van der Waals bonds.

The "hydrogen bond," found in every living thing, is an important variant of van der Waals bonding. Hydrogen atoms, with just one proton and one electron, tend to bond to just one other atom at a time—often oxygen or carbon. The lone hydrogen electron is shifted to that other atom, leaving the proton as a somewhat exposed positive bump on the resulting molecule. This positively charged region can attract other atoms by ordinary electrical force, and the bond formed in this way is called the hydrogen bond. You can picture it as a distorted hydrogen atom forming the glue that holds two other atoms together.

A molecule of water, for example, has two hydrogens attached to one oxygen, and looks very much like Mickey Mouse's head with small ears: the oxygen is Mickey's negatively charged head, while the positive protons form the ears. When solid ice forms by hydrogen bonding, the molecules all arrange themselves so that oppositely charged ends line up opposite each other. The behavior of hydrogen also explains why water dissolves so many things, from salt to sugar. Positive and negative ends of water molecules exert powerful forces on ions—ripping apart ionic crystals like table salt, ion by ion. Positive sodium ions congregate near the oxygen ends of water molecules, while negative chlorine is greeted by hydrogens.

ELECTRICAL CONDUCTIVITY AND THE CHEMICAL BOND

The most important electrical property of any material is its ability to conduct electricity—to allow electrons to flow through it. Our technological society depends on a wide range of materials with different electrical properties: efficient conductors to carry current, insulators that protect the user from harm, and semiconductors, the backbone of the microelectronics industry, which can be controlled to conduct electricity under special circumstances.

Whether a material will conduct electricity depends on how its electrons move. It is not surprising, therefore, that conductivity and chemical bonding are closely linked. Each bond type gives rise to distinctive electrical properties.

In general, materials in which electrons are loosely bound make good conductors. In such materials the presence of an outside force (supplied by a battery, for example) is enough to break the electrons loose from their moorings and start them moving. Their movement then constitutes an electrical current in the material. If, on the other hand, electrons are tightly locked in, an outside force can't dislodge them and no current will flow.

Insulators

Insulators resist the flow of electrons, and so serve such essential functions as preventing short circuits and protecting electrical consumers from shocks. Materials with ionic bonds make excellent insulators. In ionic compounds there is a one-time transfer of an electron from one ion to the next, but from then on each electron is bound tightly to a given nucleus. In this situation electrons are not easily moved. Glass and ceramics, composed primarily of ionically bonded silicon and oxygen, have long been used for the most critical high-voltage insulators.

Many covalent compounds with long chains of carbon-carbon bonds—materials known collectively as plastics—also serve as cheap, reliable insulators. Covalently bonded electrons seldom stray far from their home atom, so they are often almost as good insulators as ionic compounds. Flexible plastics, furthermore, are much easier to fabricate and use than brittle ceramics, which explains why plastic wall sockets and light switches are the norm in homes and offices.

Conductors

The terms "metal" and "electrical conductor" are almost synonymous. The electrons in a metal's negative "sea" respond easily to outside forces, and the best electrical conductors rely on this ready supply of mobile electrons. Silver is the best conductor at room temperature, but copper, almost as good, is much cheaper. Gold finds important uses as a noncorroding coating for contact surfaces, while lightweight (and cheap) aluminum is often used for carrying large currents in power lines.

Semiconductors

As the name suggests, a semiconductor is a material that will carry electricity (like a conductor), but won't carry it very well. Semiconductors walk a tightrope between a conductor with lots of mobile electrons and an insulator with none. The key to a successful commercial semiconductor is to have just a few mobile carriers of electrical charge.

The abundant element silicon provides the starting material for almost all of today's semiconductor devices. Silicon, with a half-filled outer shell of electrons, behaves very much like carbon. Silicon-silicon bonds are covalent, and pure silicon does not conduct electricity very well because the shared electrons are tightly bound. Atomic vibrations always shake some electrons loose, however, so that at any given moment there will be a few electrons wandering around in a silicon crystal. The result: silicon has a net conductivity much less than that of metals, but much more than insulators. That's why it's called a semiconductor.

The loose electrons don't tell the whole story as far as electrical current in a semiconductor is concerned. Whenever an electron shakes loose, it leaves behind a vacancy—what physicists call a hole. As far as physicists are concerned, holes can conduct electricity just as well as electrons. To see why this is so, think of the moving holes as analogous to something you see in stop-and-go traffic at rush hour. As a space (a hole) opens up in front of your car, you advance to fill that space, the next car behind moves forward to fill the new space, and so on. Most people would say the cars have moved forward, but it is just as correct to say that the hole has moved in the opposite direction. As commuters, most of us couldn't care less how fast the holes are moving, but if you are a physicist worried about the net movement of electrical charge, electrons or holes serve equally well. You can move electric charge by shifting negative electrons one way, or just as easily by shifting positive charge—the holes—in the opposite direction.

A slight impurity can alter the behavior of a semiconductor like silicon dramatically. For example, semiconductors can be made so that an atom of phosphorus replaces one out of every few million silicon atoms. Phosphorus has one more electron than silicon has in its outer shell. All but one of these electrons form the covalent bonds that hold the crystal together, leaving one electron free to wander around. Thus, the altered crystal can conduct electricity without having to depend on the electrons stolen from covalent silicon bonds. The process of adding a trace of phosphorus is called "doping" the cyrstal, and the slight excess of negative electrons yields an "n-type" (for negative charge carriers) semiconductor.

A similar situation can arise if you replace a few silicon atoms with aluminum, an element with one fewer electron in its outer shell than silicon has. Now, instead of a few extra electrons with negative charges, there are a few missing electrons—holes—with positive charge. A semiconductor doped with holes is called a "p-type" (for positive charge carriers).

Microelectronics

Microelectronics lie at the heart of all your electronic gear—TVs and radios, car ignitions and home security systems, dishwashers and pocket calculators. All of these useful things rely on semiconductor devices. In technical jargon a "device" is any composite of two or more n- and p-type semiconductors that does something useful. The simplest semiconductor device is a diode formed from two semiconductor layers, one n-type and the other p-type. When such a device is first made, electrons and holes diffuse across the boundary. If a free electron encounters a hole, it falls in and goes back to being an ordinary electron forming a covalent bond. In the process, both the hole and the free electron "disappear" in the sense that they no longer are available to move electric charge. The impurity atoms (aluminum and phosphorus in our example) for a certain space on both sides of the boundary are left without the original electrons or holes to balance their charge. The result: a layer of charged ions lines up on both sides of the boundary—positive ions on the n side and negative ions on the p side.

This double layer of charge is locked into the atomic structure of the semiconductor; once it has been formed, the charges stay where they are forever. Attracted toward the positive layer and repelled by the negative, free electrons in the material are thus accelerated across the junction. The charge layer establishes a "right" way across the junction, since it is easy for electrons to move from *n* to *p,* but hard for them to go the other way. As trivial as the junction seems to be, most of the modern microelectronics industry is based on this microscopic charge layer.

Many diodes are used as "rectifiers"—they convert alternating current of the type found in household wires to a current that flows in only one direction. While the alternating current flows in the "right" direction, the junction allows electrons to pass. When the current reverses, the electrons cannot force their way through the charge layer, so the current coming out of the device flows only one way and no longer alternates. If you look inside your home computer or TV set, chances are that the first thing encountered by alternating current from the wall is a more sophisticated version of this sort of diode. Its function is to convert the AC from the generator to DC, the type of current necessary to run the electronic gear farther downstream.

Diodes form the key ingredient in some solar energy systems, which may be important in future energy production. If sunlight strikes a thin layer of *n*-type material, it can add enough energy to jar loose some electrons. These electrons will then be accelerated across the boundary by the stationary charges that, in essence, act like an atomic-scale battery. If this so-called "photovoltaic cell" is connected to an electrical circuit, it produces a continuous supply of current so long as the sun shines. Satellites in space routinely use solar energy to run their instruments, and large-scale solar installations have been tested to see if they can substitute for conventional coal or nuclear plants. For the near term, solar power is not economically competitive with other power sources, but many scientists and engineers believe that solar energy will be a vital resource in the twenty-first century.

Perhaps the most important semiconductor device ever invented is the transistor. The simplest transistors are sandwiches formed from three slices of semiconductor—either *pnp* or *npn*—with wire leads to each of the three segments. At each boundary in the sandwich, a double layer of charge forms, just as it does in the diode. Transistors serve as fast and reliable amplifiers, detectors, or switches. They are useful because they require only a small external voltage to overcome the effects of the charge layer, thus allowing electrons to flow in the "wrong" direction across the boundary. Turn the applied voltage off and the charge layer stops the flow. In this way, a small voltage can turn a large current off and on and the transistor can act as a fast switch, a vital part of every computer. A somewhat different arrangement of applied voltages allows the transistor to act as an amplifier, taking a small current (the one from your stereo stylus, for example) and converting it into a large current (like the one that drives your speakers). There are hundreds of uses for the transistor, one of the most versatile electronic tools ever devised.

Microchips, the heart of the modern microelectronics revolution, consist of large collections of devices like the diode and transistor connected on a single piece ("chip") of silicon. Many advanced techniques can build up the necessary layers of *n*- and *p*-type domains. A common procedure is to expose the chip to a vapor containing the desired silicon-based mix and allow these materials to condense onto the surface. The so-called integrated circuits built in this way contain a complex juxtaposition of numerous *n* and *p* regions, equivalent to many transistors. Each integrated circuit is a module designed to perform a specific function, such as voltage regulation, mathematical logic functions, or timekeeping. A postage-stamp sized chip rests at the heart of your hand calculator and home computer.

Computers

Of all the microelectronic devices that engineers have produced, the computer has the greatest potential impact on society. Future historians will record that the last decades of the twentieth century marked the early stages of a revolutionary social change marked by the widespread availability of computers. Computers are everywhere—they make our airline reservations, control our traffic, ring up our supermarket purchases, and even oversee the workings of our car engines. At the heart of every computer, be it the word processor in your office or a giant supercomputer in a research laboratory, lies a microchip containing an arrangement of transistors that operate as switches—they are either "on" (allowing current to flow) or "off" (blocking the flow of current). The computer works by switching transistors from "off" to "on" and back again, millions of times per second.

The computer stores and manipulates information, which is represented by the on-off sequence of its transistors. The word processor on which this book was written, for example, stored each letter as a sequence of six on-offs, and distinguished one from another by changing the sequence. "On-on-on-on-on-on" might represent the letter "a," on-on-on-on-on-off" the letter "b," and so on.

Computer engineers measure information with a unit called the "bit," which is the amount of information stored in a single switch ("on" or "off"). A "byte" is eight bits, and typical personal computers are capable of storing and manipulating hundreds of thousands or even many millions of bits of information. The letters in this book contain about 10 million bits of information.

Superconductors

In a normal conductor, moving electrons collide with atoms, transferring energy in the process. As a result, the conductor heats up; we say it exhibits resistance to the flow of current. Superconductivity, a phenomenon exhibited at low temperature by relatively few materials, is the conduction of electricity with no resistance at all. In the world of physics, the difference between low resistance and zero resistance is all the difference in the world. Superconductors are important scientifically because they represent a state of matter quite different from normal solids. But that scientific interest is amplified by the vast technological potential of superconductors in our electronic world.

Many commercial uses of superconductors arise because they can carry large currents without heating up, and can therefore be used to make extremely powerful electromagnets. Magnets formed from traditional loops of copper wire would require literally tons of metal and rivers of cooling water to accomplish what modest-sized superconducting magnets do daily. Doctors employ these strong magnetic fields to probe the human body safely and without surgery, a technique called magnetic resonance imaging (MRI). Similar imaging equipment is used in factories and airports to test critical metal components for cracks and other defects. High-energy physicists depend on superconducting magnetics to accelerate subatomic particles so their behavior can be studied.

The so-called Maglev (for "magnetic levitation") trains are an exciting use of superconducting magnets. Such magnets in the train's body induce currents in a metal track, lifting or levitating, the train off the ground. Conventional metal-wheeled trains cannot travel much faster than 150 miles per hour, but a train that floats on a magnet cushion can "fly" at jet speeds. Japanese engineers have already demonstrated a train that will speed 300 miles per hour, fast enough to replace short-hop jet flights between nearby cities such as New York and Washington, D.C.

Superconductors also provide superfast switches for computers and extremely sensitive magnetic and radiation detectors, which are essential for specialized applications by scientists and the military. With new superconductor applications announced every month, some scientists are already referring to the twenty-first century as the "superconductor century."

The Superconductor Breakthrough

Until 1987 all known superconductors worked only at temperatures near absolute zero. Commercial superconductors were alloys of the metal niobium. Expensive liquid helium provided the necessary extreme refrigeration. The worldwide business in superconducting medical, military, and research equipment was about a billion dollars per year.

In 1987 and 1988 superconductivity made headlines when scientists discovered a new class of materials that became superconducting at much higher temperatures. New temperature records were set repeatedly, as the effect soared to a high of 125 Kelvin (about −150° C)—almost twice the temperature of cheap liquid nitrogen coolant.

The new superconductors are fascinating materials with great potential, but don't hold your breath waiting for applications. They are hard and brittle, and thus make poor wire. The materials are tricky to synthesize in pure form, are easily contaminated, and sometimes lose their superconductivity just sitting in air. In spite of these difficulties, we are confident that the new superconductors will find widespread technological uses someday. It took twenty years to develop the potential of semiconductors and lasers, so if we have to wait until the twenty-first century to enjoy new superconductor technologies it won't be too surprising.

FRONTIERS

The basic types of bonding and the resultant variety of electrical behavior have been known for decades, but new materials with special conducting properties are discovered regularly. The search for new materials with novel electrical properties remains a high priority for many researchers, and every year it seems that an important new class of materials is discovered.

New Superconductors

The search for new superconductors, which still occupies thousands of researchers around the world, will continue, though perhaps not at the frantic pace of 1987 and 1988. The "Holy Grail" of these workers is a superconductor that works at room temperature, though most physicists would probably bet against such a discovery anytime soon.

New Semiconductors

Other materials scientists will focus on the synthesis of new semiconductors and the fabrication of ever more microscopic electronic devices. Silicon dominates the industry at present, but other materials such as gallium arsenide (a one-to-one mixture of the elements gallium and arsenic) may eventually provide even faster circuitry.

Novel Conductors

Given the importance of electricity in our lives, any discovery that promises to provide new ways of using it is sure to get a warm reception from industry. The advantage to be gained from a new material may lie in the ease with which it can be manufactured and used, or it may lie in the possibility it brings of establishing a new kind of control over conducting electrons.

"Organic conductors" are made from long molecules with carbon backbones, supplemented by metal atoms. The metals turn plastics that ordinarily insulate into conductors. This introduces the possibility of building electrical circuits from plastic, an outcome that could revolutionize the electrical industry.

In another new class of materials, the atoms are arranged so that electrons cannot flow through the entire bulk of the material, but are constrained to travel in two-dimensional planes or one-dimensional channels. For example, some of these materials act as if they were made of alternating sheets of paper and tinfoil, with the electrons moving only in the latter. Others have long chains of metal atoms locked into an otherwise insulating structure. Electrons move only through the channels, and thus are confined to move in one dimension only. In both of these materials, we gain a much greater control of the electrical current than we would have with an ordinary copper wire. Engineers believe that these properties can be exploited to make whole new classes of electronic equipment.

Quantum Wires

This same sort of control can be achieved in semiconductors by devices called "quantum wires." These materials consist of thin layers of one kind of doped semiconductor sandwiched between layers of another kind. An n-type layer doped with phosphorus might, for example, be situated between two layers of n-type semiconductor doped with arsenic. These materials tend to trap electrons in the middle layer (which can be as little as one atom thick!). The sandwich is then sliced into fine strands to make the wires. At the moment this is a purely experimental technique, but it illustrates the magic of modern semiconductor technology.

Artificial Intelligence

Can computers think? Can we make a computer that is in some sense alive? These are deep questions, questions that go far beyond the boundaries of computer engineering or science itself. Unfortunately, early investigators in the field of artificial intelligence promulgated a wildly optimistic point of view that has given the whole field something of a bad name. This hype has caused some commentators to overlook real progress in recent years.

There are two approaches to artificial intelligence. One is to use the computer's speed and power to overcome obstacles. A computer chess program, for example, tries to play all possible future games to choose its next move. No human plays this way, of course, which probably explains why the best human players routinely defeat the best machines. More recently, the emphasis has shifted to trying to design machines and programs that imitate human thought. Researchers try to understand how the human mind operates, how it reaches decisions on the basis of partial or fuzzy information, how intuition mixes with reason. This work has already resulted in important applications like "expert systems" (machines that store and manipulate knowledge in imitation of human experts) and pattern recognition.

Work in this new approach to artificial intelligence has just begun, and it is difficult to tell how far it will carry us toward a machine that we can characterize as alive or intelligent. For the record, we doubt that any machine will ever be able to produce the sorts of things we identify as most quintessentially human — a play like *Macbeth,* a theory like general relativity, or a Mozart sonata.

CHAPTER 7

ATOMIC ARCHITECTURE

Think again about the following list of things:

an elephant
pantyhose
the Empire State Building
sand
your left ear
the Pacific Ocean
air
tofu
Jupiter
beer
this book

If they're all made of just a few different kinds of atoms, how come they're so different?

The answer can be found as close as the "lead" in your pencil and the diamond ring on your finger. It's hard to imagine two solids more different than pencil lead (also known as graphite) and diamond. One is black, the other colorless. One is so soft it leaves a streak on paper, the other so hard it scratches anything. One has a flat, dull surface, the other shines with a brilliant luster. Graphite costs pennies a pound, diamonds can cost millions. Yet both graphite and diamond are pure forms of the element carbon. The only difference between them is the way nature packed the carbon atoms together. Studies of atomic architecture in graphite, diamond, and tens of thousands of other materials reveal that:

The way a material behaves depends on how its atoms are arranged.

Everything you touch or feel, all the objects of our world with their extraordinary range of appearances and properties, consist of different arrangements of atoms. There are only a few dozen different kinds of common atoms, but there are infinite ways to organize them into gases, liquids, and solids.

THE STATES OF MATTER

Most of our experience with atoms has to do with atoms in combination, rather than taken singly. A clump of atoms bound together is called a molecule, and molecules form our familiar world. What you wear, what you drive, even what you eat and breathe, all are made from them. Depending on how the atoms are organized, large-scale collections of atoms or molecules may appear quite different from each other, even though they contain the same set of atoms. Scientists call each style of organization a "state of matter," and the three most familiar are gases, liquids, and solids.

Gases

Gases fill balloons, propel bullets, and form Earth's atmosphere. Many gases like the air you breathe are invisible, but you know something's there when the wind blows. The distinctive feature of all gases is their ability to expand, filling whatever volume is available. This behavior reflects the atomic structure of gas.

Each gas particle is an individual atom, like the gaseous elements neon or helium, or a molecule with two or more linked atoms, such as oxygen (O_2), carbon dioxide (CO_2), or methane (CH_4). If we could magnify gas atoms and molecules a hundred million times they would look something like the wildly flying

Ping-Pong balls of state lottery games. Gas particles do not stick to one another, but rather careen off walls and collide with other particles. The gas pressure that fills your car's tires is a consequence of the impact forces of countless trillions of these speeding projectiles, bouncing off each other and the inside of their confining vessel. If you crowd more particles into a fixed space like your pressure cooker, or if you reduce the volume as when you squeeze on a balloon, then pressure must increase. Heating gas, which causes molecules to move faster and collide with more force, also increases the pressure. That's how your pressure cooker works.

The energy of chemical explosions results when dense solids or liquids are quickly converted into hot expanding gas. Bullets, cannonballs, and rockets accelerate because of this force, which is nothing more than the cumulative effect of individual atomic collisions.

At extremely high temperatures, like those in the sun, gas takes on a different character. In this gas-like state of matter, called plasma, electrons are stripped off atoms. At low temperatures, only a few electrons per atom may be separated, but at very high temperature—typically above 100,000 degrees—electrons are completely stripped from gas molecules, yielding a complete plasma. Plasmas exhibit unusual properties not encountered in typical gases. For example, plasmas conduct electricity and they can be confined in magnetic fields called magnetic bottles.

Even though you've never actually felt a plasma, they are by far the commonest state of matter in the universe. Every star, the sun included, is composed primarily of dense hydrogen- and helium-rich plasma, while tenuous plasma-like gas occurs in the outer atmospheres of several planets, including Earth. There is even plasma (though one in which only a few electrons have been stripped from each atom) in your fluorescent light bulb.

Liquids

Like gases, liquids have no fixed shape, but they differ from gases in having a fixed volume. At the atomic level liquids behave something like a bowl full of marbles. Like marbles, the liquid molecules slide over each other, easily shifting to fill any available volume. Individual molecules do not stick tightly together, however, so the entire mass is free to change shape (or spill on the floor).

Solids

Solids are materials that have a more or less fixed shape. Atoms in solids bind together with sufficient strength to stay in place. Crystals, glasses, and plastics, three of the most important varieties of solids, differ primarily in the regularity of their atomic structures.

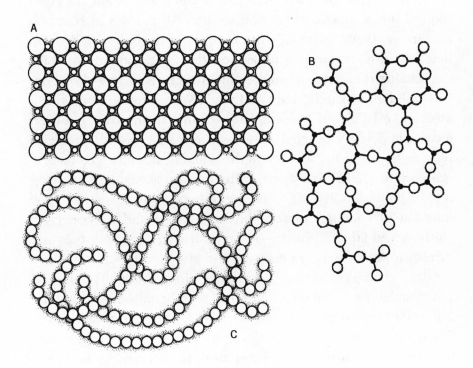

Three different kinds of solids are distinguished by the regularity of their atomic structures. Crystals (A) possess row after row of identical atom boxes; trillions of boxes stack to form one tiny crystal. Glasses (B) have random structures; there is no way to predict what atom might reside a short distance from any starting point. Plastics (C) form from long chains of carbon atoms called polymers; atoms are ordered in one direction (along the chain) but are spaced randomly elsewhere.

Metals, gemstones, bones, and computer chips are crystals. They contain regular three-dimensional arrays of atoms, repeated over and over again in a sort of Tinkertoy arrangement. You can imagine a crystal as something like a huge stack of boxes, with each box the same size and shape and all the boxes holding exactly the same atomic contents. The size of each "box" in a real crystal is usually no more than a ten millionth of an inch, and each box might contain up to a few dozen atoms. The structure of a crystal is orderly, with row upon row of exactly the same kinds of atoms; travel 100 boxes in any direction and you know exactly what you'll find.

Plastics, invented in the twentieth century, are among the commonest solids we encounter in daily life. All plastics in common use are synthetic chemicals not found in nature, but built from long, chain-like molecules linked together by carbon atoms. Like any chain, they have a predictable structure in one direction; travel along any single chain and you'll keep running into carbon atoms like beads on a string. The chains themselves, however, form a hopelessly tangled mass of interwoven strands that give plastics their rigidity and strength. When heated, the chains separate slightly and slide apart, softening the plastic and allowing it to be poured, rolled, or otherwise reshaped into new forms. The long list of common plastics includes nylon and polyester (in clothing and fabrics), Lucite (in sculptures), vinyl (floor tiles and furniture), and epoxy (in cements and glue).

Glass structures differ considerably from both crystals and plastics. Atoms in a glass are jumbled together almost at random, but usually with some regularities in the atomic architecture. Common window glass, for example, is made from silicon and oxygen, with almost every silicon atom surrounded by four oxygen atoms and almost every oxygen atom next to two silicons. But there aren't any "boxes" of structure neatly stacked in window glass. Travel just a few atoms in any direction from any point and there is no way to guess whether you will find an oxygen or a silicon atom. Beyond the nearest neighbor atoms, the structure is random.

Changes of Phase

If an ice cube falls on the floor it melts, leaving a puddle for you to clean up. If you boil water on the stove, you produce a cloud of steam. If you let your paintbrush dry without cleaning it, the brush becomes hard and useless. These are examples of processes in which matter changes phase—from solid to liquid in the first case, from liquid to gas in the second, and from liquid to solid in the third. The atoms and molecules are the same before and after the change of phase, but their relationship to one another is different.

In the ice cube, for example, the molecules of water in the crystal lattice vibrate faster and faster as heat from the outside air moves in. Eventually they vibrate so fast that they tear loose and begin to move around freely. When this happens, the ice turns to water and we say that the cube has melted. In just the same way, molecules in boiling water move faster and faster until they lose contact with one another and enter the air as a gas. As paint hardens, on the other hand, atoms link up to form long chains where before there were separate molecules. Physicists know melting, boiling, and solidification as "phase transitions."

All phase transitions require energy. Materials soak up heat while they melt or boil, but their temperature stays the same—all of the energy input goes into breaking up the old atomic arrangement. This is why you use ice cubes to keep your drink cold. When you take an ice cube from the refrigerator, it is at a temperature well below freezing. It starts to warm up as it absorbs heat from your drink, cooling the drink as it does so. When the ice gets to 32° F, however, it remains at that temperature until it has absorbed enough energy to break all the bonds between the water molecules in its crystal structure. Only after the cube has melted completely and all the water is in liquid form does the water temperature start to rise again.

PHYSICAL PROPERTIES

When you go to the grocery store you test the food you're about to buy. Are the tomatoes firm? Does the meat have good color? Is the lettuce crisp, the bread soft? As you evaluate your potential purchases you are determining physical properties of the materials you buy.

A physical property is any aspect of a material that you can measure. Every measurement has three essential components: a sample, a source of energy, and a detector. The sample is the thing you want to measure. Samples vary widely in size and character—you can study single subatomic particles, a piece of fruit, the whole earth, entire galaxies, and anything in between.

All measurements depend on an interaction of the sample with some form of energy. Color is the interaction of matter with light. Electrical conductivity is the interaction of matter with an electric field. Brittleness can be measured by the interaction of matter with the force of a moving hammer. Without energy, you can't measure these properties.

The detector measures the interaction between the sample and the energy. The human senses provide marvelously sensitive and portable detectors such as the eyes and ears, and scientists have developed many detectors that extend our senses, such as photographic film, speedometers, thermometers, and radios. All of these detectors (and many others) record the interactions of matter and energy.

We are surrounded by countless materials, each with properties ideally suited to its function. You are reading a book with thin, white, flexible paper pages, fast-drying black ink, and strong glues for binding. You are probably sitting in a comfortable chair constructed from some combination of laminated woods, resilient plastics, light metal alloys, and synthetic fabrics. We wrote this book on a word processor made of semiconductor integrated circuits, a glass vacuum tube display screen, flexible copper wires, and colorful plastic insulation.

All of these adjectives—flexible, strong, light, colorful—refer to physical properties that you can measure. Scientists organize the dozens of different physical characteristics into a few basic categories, including mechanical, magnetic, and optical properties. All of these properties are determined by the atoms and how they fit together.

MECHANICAL PROPERTIES

Mechanical properties include a host of characteristics that describe how a material withstands stress and strain. Does it scratch easily? Does it break when you squeeze, twist, or stretch it? Mechanical properties are critical to thousands of day-to-day applications. You want your floor to be strong, your bed soft, and your toothbrush bristles flexible. Thousands of materials scientists spend their professional lives looking for "new and improved" products with useful mechanical properties.

Elasticity and Strength

Some materials are strong and others weak; some boughs bend while others break. These profound differences in material properties reflect equally profound differences in atomic structures. When two objects, like your feet and the floor, come into contact, two forces are involved. The force of gravity pushes you down, but an equal and opposite electrostatic force acts as electrons in feet and floor try to occupy the same space. Any time two chunks of matter are forced together, both objects must experience some deformation, even if only a subtle one. (Your feet become a little flatter, for example, and the floor bends just a bit.) Under mild stress most solids like feet and floors deform elastically; when stress is released they spring right back to their original shape. If too much force is applied, however, materials reach their "elastic limit" and a permanent deformation occurs: metals bend, paper rips, glass shatters, fruit squishes.

Elasticity and strength have their origins in atomic structure. We already have part of the story in the relative strength of different chemical bonds. Clearly a material can't be strong if all its bonds are weak. But even compounds with the strongest bonds can be weak if the atoms aren't properly arranged. The way bonds are assembled into a three-dimensional structure makes all the difference. To grasp this point, consider again the two forms of pure carbon, diamond and graphite. Diamond is the hardest, strongest material known. In a diamond crystal, strong covalent bonds link each carbon atom to four neighbors. The entire carbon array forms a rigid three-dimensional framework.

Carbon-carbon bonds also dominate the graphite structure, but each atom is tightly bound to only three neighbors in layers of atoms. Graphite layers are held together by van der Waals bonds so weak that graphite is an effective lubricant and one of the softest known minerals. Every time you use a "lead" pencil,

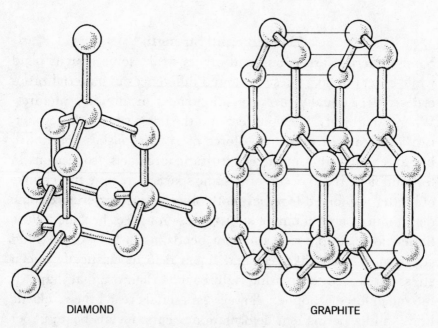

DIAMOND   GRAPHITE

Diamond and graphite are both crystals formed entirely from carbon, but their atomic structures are strikingly different. In diamond, each carbon atom is surrounded by four neighbors in a rigid three-dimensional framework. Graphite, on the other hand, is a layer structure with each carbon atom near only three others.

black layers of graphite shear off the end and transfer to paper because the small force you exert on the pencil is enough to break the van der Waals bond. Graphite illustrates the common addage that a chain is only as strong as its weakest link. The distribution of strong and weak bonds, rather than the type of atom, makes the critical difference in mechanical properties.

Carbon-carbon bonds hold together the strongest known fibers. Long chains of carbon atoms are essential structural elements of spiders' webs, nylon fibers, wood, and a host of plastics. These chain molecules, or polymers, can also make the best elastic bands when carbon-carbon chains adopt a folded or zigzag form, straightening only under tension.

Ductility and Brittleness

Grab a hammer and strike a piece of fine china with all your might. The ceramic shatters into a thousand pieces, each retaining the shape of a fragment of the original dish. Do the same thing to a lump of lead. The lead deforms, absorbing the blow by changing shape. These differences reflect the flexibility of atomic bonding.

China has ionic bonds, with alternating plus and minus ions that cause inflexible links in fixed directions. Attractive forces line up exactly along atomic pairs. Ionic crystals are like models constructed of rigid balls and sticks. Once a few sticks come loose the results are often catastrophic. If bent too far, the sticks break and are not easily reformed.

Bonds in metals, with their freely roving electrons, are much less directional. Metal atoms are something like marbles in a bowl of molasses. Each round marble fits neatly into the array, surrounded by its neighbors. Molasses, like the metal's electron sea, keeps everything stuck together. Tilt the bowl, however, and the marbles slowly slide past one another, adopting a new arrangement. The efficient packing of marbles is retained, but individual marbles shift. The same sort of thing happens when a metal is hammered or bent. Layers of atoms slide against each other, yielding a new shape to the metal object, but the bonding is preserved.

Rubber balls adopt a different strategy in response to a blow. Individual covalent carbon-carbon bonds in rubber are strong, but they have a great deal of directional leeway. The stress of a bat or foot deforms the ball, forcing the atoms closer together momentarily. This process parallels what happens when you squeeze a spring: you do work, adding energy to the system, which is stored as elastic potential energy in the spring. But though the bonds may bend, they do not break. The stored energy reconverts to kinetic energy as the bonds straighten out and the ball snaps back into shape and zooms off. Most sports would be a lot less interesting if we only had metal or ceramic balls!

Composite Materials

If you've performed any sort of do-it-yourself work around your home you've probably run into composite materials, one of the hallmarks of modern materials science. Composites overcome the shortcomings of one substance by combining it with others. Shatterproof car windshields, integrated circuits, and steel-reinforced concrete are all examples of the new technology.

Plywood is the classic composite material. It consists of thin layers of wood glued and pressed together, with the grain direction alternating layer by layer. The resulting lumber has no weak direction due to the grain, so plywood is stronger than traditional boarding of the same thickness. Furthermore, it can be fabricated in large sheets from relatively small trees, because the wood veneers can be sliced off a rotating tree like paper towels off a roll.

Fiber composites, in which strong, flexible strands are embedded in epoxy resins, provide lightweight materials of unusual strength. Fiberglass, made of hair-like glass threads woven, molded, and impregnated with a resin cement, is still the most widespread of these materials. Engineers increasingly use carbon fiber composites in aircraft design, and they provide an extra punch to tennis rackets and golf clubs.

MAGNETIC PROPERTIES

All magnetic fields are created by moving electric charge, but when you look at a refrigerator magnet it's not obvious that anything is moving. The key to the magnetic properties of refrigerator decorations, like all other material properties, lies at the atomic scale.

Every electron revolves around the atomic nucleus. If you could stand at one point along the orbit and count, you would see an electron go by every so often. The single moving electron, then, is a tiny electric current. These orbiting electrons, like other electric currents, create magnetic fields, and their net effect is to provide the atom with a magnetic field as well. This means that you can imagine replacing each atom in a material by a tiny bar magnet with its own north and south pole. In the vast majority of materials these atomic magnets point in random directions, and the magnetic fields associated with them cancel each other out, so they can't exert a magnetic force.

In a few materials, notably some compounds of iron, nickel, and cobalt, the orientations of atoms and their orbiting electrons are not entirely random. In these materials the atomic magnets line up with each other. This alignment takes place in blocks several thousand atoms on a side, called "domains." Within each domain, each tiny atomic magnetic field reinforces the whole. In a permanent magnet neighboring domains tend to line up and reinforce each other as well, and the material as a whole is capable of exerting a large magnetic field. If the material is heated, however, the alignment of the domains can be scrambled and the magnet reverts to being an ordinary piece of iron, even though a uniform alignment is retained in each domain.

The atomic-scale origin of iron's magnetism explains why every permanent magnet has a north pole and a south pole and why no isolated magnetic poles exist. If you break apart a magnet you always get small pieces with north and south poles. We can now see that this result follows from the fact that an electron in orbit is like a loop of electrical current, and that such loops always produce dipole fields. This statement is valid down to the basic unit of magnetism, the atom. But take an atom apart and the magnet disappears.

OPTICAL PROPERTIES

We humans have a special interest in the way matter behaves when it meets the very narrow range of electromagnetic radiation called visible light. There is nothing intrinsically different about visible light compared to any other part of the spectrum; we could just as easily talk about the interaction of matter with radio or gamma rays. But visible light is special to us because we have a pair of sophisticated light detectors—our eyes—that allow us to determine luster, transparency, color, and a variety of other optical properties.

The sensitivity of our eyes to red, yellow, green, and blue is not pure chance. The sun's energy output peaks around visible light, and Earth's atmosphere is transparent to those wavelengths. Coincidentally, the energy of light is similar to that of many kinds of bonding electrons. When you observe colors and lusters, you are, in effect, probing the world of atomic bonding.

Light interacts with matter in several ways. Some light waves pass right through a crystal without any effect more noticeable than tides have on a ship at sea. Other waves do interact with atoms, but are absorbed and reemitted unchanged. The net result of these two processes is that light energy can travel through matter, a process we call transmission. Window glass, water, and the atmosphere all transmit light.

Light moves a little more slowly in matter (with atoms) than in a vacuum (with nothing), so transmitted light waves undergo a slight change of direction at the surface, a phenomenon called refraction. Familiar effects like the shimmering of air above a heated highway, the apparent distortion of someone standing in a swimming pool, and the concentration of light by a lens result from refraction. Dozens of everyday optical devices, from microscopes and telescopes to eyeglasses and cameras, rely on lenses that bend light rays to achieve their aims.

Some light waves bounce off solids like ripples off the side of a boat, a phenomenon called scattering or reflection. The angle at which light strikes a flat surface exactly matches the reflected angle. Most surfaces are rough, however, so that each light wave scatters through a slightly different angle and no image is formed. On very smooth surfaces, such as polished metals or glass in a mirror, the scattering is more uniform.

Other light waves add energy to the crystal and disappear—the process of absorption. Ordinary sunlight (or white light) is a mixture of light at all wavelengths—or all colors. A material we perceive to be colored absorbs some visible wavelengths more than others. A leaf that absorbs red light looks green to us, for example, and a stained-glass window that absorbs blue looks orange. If an object absorbs all the wavelengths, we call it black.

Each tiny packet of light—each photon—has a choice. It can be transmitted, reflected, or absorbed. Each of these events is a different kind of interaction between light and atoms, and not all photons have to do the same thing. When you stand in front of a shop window you see your ghostly reflection, proving that some light is being reflected. But at the same time you can view the goods on display as some light passes through the glass and back out again. Water transmits most light, but dive below about 100 feet and things get pretty dark because each foot of water absorbs a small fraction of the light that comes to it.

Opaque colored materials, like dyed clothing, poster paper, or flowers, scatter the light they don't absorb. Transparent colored materials, including stained glass, gemstones, and mixed drinks, transmit the unabsorbed light rays. Your body absorbs some light, but reflects some light both from the surface and from a fraction of an inch below the surface, thus giving healthy skin a kind of "glow." Combinations of transmission, absorption, and scattering result in many distinctive surface lusters: greasy, metallic, waxy, or dull. For each visible wavelength, any fraction of photons may be transmitted, absorbed, or scattered, leading to an infinite variety of colors and lusters.

The question of why a particular material reacts with light in a particular way requires a complicated explanation, but it always depends on the way atoms and their electrons are linked. You can think of the process leading to color in this way: an incoming lightwave shakes an electron, accelerating it much as an ocean wave jostles a piece of driftwood. The accelerated electron in turn converts the energy it has received into another electromagnetic wave, which radiates away from the electron like radio signals from an antenna. The amount of original energy absorbed or radiated depends on the web of forces that holds the electron in place, and hence is extremely sensitive to small changes in the composition of the material.

Bright Colors

It doesn't take much to create vivid colors. Rubies, emeralds, and sapphires, the most prized of all colored gems, are all varieties of common colorless minerals. Adding a few atoms per thousand imparts a deep color (and high value) to otherwise bland rocks. A deep red ruby is nothing more than common aluminum oxide spiced with a trace of the element chromium (an element best known for its use in shiny chrome automobile bumpers). Chromium in a ruby soaks up a broad band of green energies; our eyes and brain interpret this lack of green as red. Most of the bright colors of objects you see around you arise from the same type of absorption. We "see" a color that is complementary to the one being absorbed.

In special instances, bright colors can arise from a different process. In so-called fluorescent materials, electrons absorb energetic radiation (commonly ultraviolet or "black light" radiation), raising an electron to a higher Bohr orbit. The energized electron soon drops down in a series of steps through lower energy states, and in the process releases one or more photons at lower energy corresponding to visible light. These emitted photons give the material a vivid, bright color because they are concentrated in one very narrow range of wavelengths. When you look at a red neon light, fluorescent paint, or Day-Glo colors, you are seeing such a narrow band of photons.

FRONTIERS

Phase Transitions

Melting and boiling aren't the only kinds of phase transitions in nature. When a magnet forms in hot iron or a piece of cold metal becomes superconducting, more complex kinds of phase transitions occur. The study of these transitions occupies many theoretical physicists these days. They are finding that as different as these various phase transitions may appear to be, there are simple underlying regularities common to all of them. This suggests that when matter becomes more ordered (as in freezing) or more disordered (as in melting), it does so according to some as yet unknown general law. If we knew this law, we could understand exactly how it is that atoms establish themselves in a given order in materials.

The Search for New Materials

Scientists look for patterns in nature—patterns to help us understand our universe and shape our environment. Countless thousands of materials have been synthesized and studied by materials researchers, who see again and again that atomic architecture dictates physical properties. If you want a strong fiber, use chains of carbon atoms. If you want a flexible electrical conductor, stick with metals. If you want a tough electrical insulator, find a ceramic. We can use our understanding to design—atom by atom— new paints, new plastics, and new materials for golf clubs, window glass, and computer chips.

Look around you. Most of the objects you use in your home, at work, and school are the result of materials research. No one can predict the future, but you can be sure that in the coming months and years, scientists who engineer atoms will discover new materials that will change our world.

CHAPTER 8

NUCLEAR PHYSICS

If you are like most people, you're probably pretty suspicious of nuclear physics. You may immediately think of a Tom Clancy novel with silent machines of destruction lying in wait deep beneath the world's great oceans. You know that submarines, driven by powerful nuclear reactors and stocked with dozens of atomic weapons, are poised to unleash their doomsday arsenal.

It is certainly true that nuclear submarines control tremendous potential energy. A few pounds of nuclear fuel provide enough energy for a submarine to travel around the world and sustain a crew for months without surfacing. Nuclear warheads, each no larger than a bag of groceries, have the power to destroy cities. But the energy of the nucleus reveals itself in much more than weapons. In every major hospital, particles emitted by nuclei are used to destroy cancerous cells. Other nuclei, when injected into the body, allow physicians to trace the movement of atoms and make diagnoses. In the natural world, the sunlight that nourishes our planet comes from nuclear reactions inside the sun.

But whatever the use to which nuclear energy is put, it can always be understood in terms of a simple basic principle:

Nuclear energy comes from the conversion of mass.

We now realize that mass is a very concentrated form of energy. The nucleus of the atom, dense and heavy, accounts for most of the atom's weight and almost none of its volume. The nucleus, which is responsible for nuclear energy and radioactivity, behaves independently of the atom's far-flung electrons, which control chemical bonding. The nuclei of most familiar atoms are stable and do not change, but others disintegrate and send out energetic particles that we call radiation. In these reactions, some of the original mass converts to energy. Energy can also result from nuclear reactions like fission (the splitting of the nucleus) and fusion (the coming together of nuclei). In both cases, the energy obtained from a nucleus comes from the conversion of mass into energy.

THE NUCLEUS

The atom is almost all empty space. If the nucleus of a uranium atom were a bowling ball sitting in front of you right now, the electrons in orbit would be like 92 grains of sand scattered over an area equal to that of a good-sized city. Yet despite the small size of the nucleus, it ties up virtually all of the mass of an atom. In a rough way, you can say that the electrons determine the size of the atom, while the nucleus determines its weight. With so much mass crammed into such a tiny volume, the energy locked up in the nucleus is immense. This is why an atomic bomb, which works by rearranging things inside of nuclei, is so much more destructive than ordinary chemical explosives, which work by rearranging electrons in their orbits.

The difference in size between the nucleus and the atom also helps explain an important characteristic of matter: what happens in the nucleus is largely independent of what happens to the electrons and vice versa. The electron in the suburbs doesn't really care what the nucleus in the city is doing, and vice versa. Since chemistry involves the outer electrons, chemical reactions do not depend to any great extent on what happens in the nucleus. Similarly, the actions of the distant electrons do not affect what happens in the nucleus. These facts have extraordinary practical consequences.

NUCLEAR ENERGY

The two principal kinds of nuclear particles, protons and neutrons, are locked tightly into the structure of the nucleus. It takes tremendous amounts of energy to change this nuclear structure. In the outer regions of an atom, electrons emit visible light when they move from one orbit to another. Inside the nucleus, a proton or neutron making a similar change emits an X ray with a million times as much energy as that contained in visible light. The energy available in the nucleus is much higher than that available in the rest of the atom.

Almost all nuclear energy comes from conversion of mass. The mass of a nucleus is typically somewhat less than the sum of the masses of the protons and neutrons that would be assembled to create it. The nucleus of carbon-12, for example, has a mass about 1 percent less than the mass of six protons and six neutrons. When a carbon-12 nucleus forms, the excess mass is converted into energy via the formula $E = mc^2$, and this energy holds the nucleus together.

There are two ways to tap the energy of the nucleus: fission and fusion. In both cases, the energy we get comes from the conversion of mass into energy, and in both cases the energy is available because the mass of the final state of the nuclear system is less than that of the original.

Fission

A nucleus undergoes fission when it splits into two or more fragments. Usually the combined masses of the fragments is greater than the mass of the original nucleus. You normally have to put energy into splitting the nucleus, just as a logger puts energy into swinging an ax to split a log. Occasionally, however, the sum of the masses of the fragments is less than the mass of the original nucleus. In this case, the fission releases energy that is usually called "nuclear energy."

The most familiar nucleus that yields energy when it splits is an isotope of uranium called uranium-235 (92 protons, 143 neutrons). This isotope makes up less than 1 percent of naturally occurring uranium (the most common form is uranium-238). If a slow-moving neutron collides with uranium-235, the nucleus splits into two roughly equal fragments and two or three more neutrons. The sum of these masses is less than that of the original nucleus, and the difference in mass is converted into the energy of motion of the fragments. This energy, eventually released as heat, is used to run commercial nuclear reactors and generate electricity—perhaps even the electricity that provides the light you're using to read this book. As you read, atoms of uranium are dying to provide your light.

The heart of a nuclear reactor is the core, a large stainless steel vessel that holds several hundred fuel rods. These pencil-thin shafts of uranium, rich in the isotope uranium-235, are separated by a fluid (usually water) whose atoms collide with neutrons and slow them down. When a fission occurs in one fuel rod, the fast neutrons leave the rod, are slowed down in the water, and then

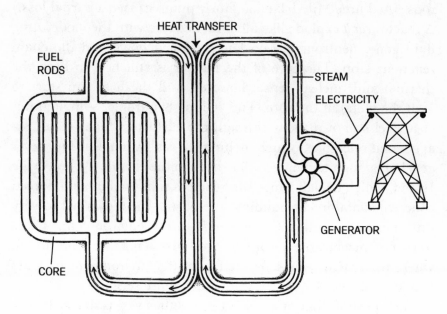

Nuclear reactors contain a core with radioactive nuclear fuel rods. Nuclear chain reactions in this fuel heat a surrounding jacket of water that converts heat into steam. The steam, in turn, drives an electric generator.

enter another rod and cause more fissions. These fissions each produce two or three neutrons of their own, each of which can go on to cause still more fissions in other rods. Scientists call this proliferation of collisions a chain reaction, the rate of which is regulated by lowering control rods of neutron-absorbing materials down between the fuel rods.

The energy of the fission fragments heats the water, which is pumped out of the core. In another part of the generating plant this heat produces steam, which runs a conventional generator and produces electricity. Thus, a nuclear reactor differs from a coal- or oil-powered plant only in the way it produces heat. Once the steam is made, everything else is the same.

The aspect of nuclear reactor operation that most often occupies public attention is the possibility of accidents. The names Three Mile Island and Chernobyl conjure up visions of radioactive nightmares. The most serious reactor accidents (which are also the most unlikely) involve loss of the fluid that separates fuel rods. (At Three Mile Island, a faulty pump caused a partial loss.) A reactor *can't* explode like a bomb, because with the moderating fluid gone, neutrons are no longer slowed down and the chain reactions stop. The core of the reactor is still hot, in both the thermal and nuclear sense, however, and the heat can start to melt the metal in the core. The "China Syndrome"—molten nuclear fuel so hot it melts through the earth to China—is an exaggerated reference to such melting. In reality, nuclear fuel never gets hot enough to melt very far through the earth. At Three Mile Island, which like all American reactors is housed in a reinforced concrete containment building, the partial meltdown led to very little escape of radioactive material to the environment. There were no measurable public health problems. At Chernobyl, where the reactor was separated from the environment by glass windows, the consequences were much more serious.

The question that faces us all is whether we, as a society, are prepared to accept the (admittedly small) risks associated with nuclear power to gain the advantages of electricity generated by reactors. This isn't a scientific question, but one of values—of weighing costs versus benefits. But in order to make a decision, every citizen should know some basic facts about reactors and radioactivity.

Fusion

Fusion occurs when two small nuclei come together to form a single larger one. As with fission, it sometimes happens that the mass of the final fusion product is less than the mass of the ingredients that go to make it up. In this case, the fusion process can produce energy. The sun and other stars generate their energy through fusion in a process in which four protons (the nuclei of hydrogen atoms) come together after a sequence of steps to form a helium nucleus and a few extra particles.

Since the 1950s there have been extensive research and development programs in both U.S. and foreign laboratories to harness fusion as a source of electrical power. The general strategy is to hold nuclei together with intense magnetic fields, then raise their temperature to reproduce conditions that occur in the interior of stars. Unfortunately, this program remains far from producing controlled fusion in a laboratory, much less an economically viable generating plant. There have been periodic announcements of breakthroughs over the past thirty years, but fusion power still remains a dream that will be realized only in the mid-twenty-first century, if ever.

In the spring of 1989, a brief flurry of excitement erupted when two scientists at the University of Utah claimed to have produced fusion in a desktop experiment. Dubbed "cold fusion" or "fusion in a bottle," this process caught the imagination of the public because it might have provided us with an unlimited cheap source of energy. But cold fusion has faded from view as other scientists have been unable to reproduce the claimed results.

Nuclear Weapons

Nuclear chain reactions can run out of control. Take 25 pounds of uranium-235 and the mass will just sit there, giving off heat and spewing out neutrons. But place two 25-pound masses together and you can produce streams of neutrons that multiply into an unstoppable torrent—a nuclear explosion. The atom bomb uses this principle by separating two precisely machined hemispheres of uranium-235 and surrounding them with conventional chemical explosives. The first explosion pushes the hemispheres together into a single sphere that exceeds the "critical mass."

The fusion of hydrogen into helium produces even more explosive power—in the hydrogen bomb. H-bombs are triggered by atomic bombs, which provide the heat and compression necessary to begin the fusion reaction. Hydrogen bombs thus unleash the same kind of energy as the sun. Atomic bombs can't be much larger than a critical mass, but there is almost no limit to the size of a hydrogen bomb. The more hydrogen explosive you start with, the larger the bang.

RADIOACTIVITY

Most familiar nuclei are stable. Almost all the nuclei of carbon in your tissues and calcium in your bones are the same now as they were when they were made in the heart of a supernova billions of years ago. Some nuclei, however, do not share this property. In periods ranging from microseconds to times comparable to the age of the Earth, these nuclei spontaneously disintegrate, spewing out fragments when they do so. These nuclei are said to be radioactive, the process of disintegration is called "radioactive decay," and the particles emitted during this decay constitute radioactivity. All isotopes of uranium are radioactive, as are many lighter nuclei such as carbon-14 and strontium-90.

Half-life

The best way to think about the behavior of radioactive nuclei is to picture popcorn popping on your stove. Kernels don't all pop at once. A few kernels pop, then a few more, spaced out over several minutes. No theory of nuclear stability exists that will always predict all the details of radioactive decay. We don't understand, for example, why some atoms take billions of years to decay while others disappear in a matter of seconds. However, we can measure the phenomenon with great precision.

Any given collection of radioactive nuclei behaves in roughly the same way as another, with individual nuclei decaying at different times. The overall rate of decay is called the half-life, defined as the time it takes for half of the nuclei in a given sample to decay. This means, for example, that if you have 100 nuclei with a half-life of one minute in front of you right now, you will have about 50 one minute from now, about 25 (half of a half) in two minutes, 12.5 (on the average) in three minutes, and so on.

Half-lives of nuclei vary widely. Uranium-238 (the most common isotope of uranium) has a half-life of 4.5 billion years — about the same as the lifetime of the earth. The shortest-lived of plutonium's many isotopes, on the other hand, has a half-life of a billionth of a second, so its decay can be measured only with sophisticated electronic detectors. A range of values between these two extremes exists in nature.

Alpha, Beta, and Gamma Decay

Radiation, first discovered at the end of the nineteenth century, was mystifying to classically trained physicists and chemists. There seemed to be three different types of radiation, each coming from a different mode of decay. Scientists named these mysterious types of radiation for the first three letters of the Greek alphabet — alpha, beta, and gamma. We still use those names today, even though we understand a great deal more about all three types.

When a nucleus undergoes alpha decay, it emits a bundle consisting of two protons and two neutrons — the nucleus of a helium atom, also called the alpha particle. After alpha decay, the "daughter" nucleus has two fewer protons and two fewer neutrons than it had originally. This means that the nucleus can attract two fewer electrons with its electrical force, and after a time the two excess electrons wander off. What remains, then, is an atom with two fewer protons and two fewer electrons; an atom of a different chemical species has been created. Thus alpha decay changes both the mass and the identity of the nucleus involved.

Uranium-238, for example, decays by emitting an alpha particle, and the end product is an atom of the element thorium (thorium-234, to be exact). Alpha decay, by changing the identity of the nucleus, changes the identity of the atom itself. Alpha decay (and, as we shall see shortly, beta decay) constitutes a modern version of the philosopher's stone, the material medieval alchemists believed could change lead into gold.

In beta decay, one of the neutrons in the nucleus emits an electron and, in the process, converts itself into a proton. The daughter nucleus has almost the same mass as its parent, but has one more proton and one fewer neutron. Beta decay, then, changes the identity, but not the mass, of a nucleus. The "beta particle" was named before people realized that it was a plain old garden-variety electron, and you still see electrons referred to occasionally as "beta rays."

One of the most intriguing beta decays doesn't involve a nucleus at all, but a free neutron. Left to itself, a neutron decays into a proton, an electron, and a particle called the neutrino with a half-life of about 8 minutes. Neutrons in a nucleus can't normally decay in this way. Thus we still have neutrons around, billions of years after the creation of the universe, only because they have been hiding inside of nuclei.

Finally, gamma decay involves a rearrangement of protons and neutrons inside the nucleus and the consequent emission of electromagnetic radiation in the form of an X ray. Gamma decay changes neither the mass nor the identity of the nucleus.

Decay Chains

The story of radioactivity does not usually end with one decay. Typically, a radioactive nucleus decays by one process, producing a daughter that decays by another. The daughter of the second decay will decay in turn, and the process goes on through a long chain until it ends with a stable nucleus. This process is well illustrated by uranium-238, a surprisingly common element in the Earth's crust (much more common than gold, silver, or mercury, for example). It decays by alpha emission to thorium-234, which decays by beta emission to protactinium-234 (91 protons, 123 neutrons) with a half-life of 24 days. This nucleus in turn decays by beta emission to uranium-234 with a half-life of 2 minutes, and uranium-234 emits an alpha particle to create thorium-230 with a half-life of 250,000 years. This series of decays continue, until it ends with the formation of lead-208, a stable nucleus.

One inevitable product of the decay chain that starts with uranium-238 is the radioactive gas radon-222. Radon decays by alpha emission with a half-life of about 4 days. This gas can seep up from the ground into your home, where its decay products may pose a significant health risk. Taking the long view, we recognize that the indoor radon problem is caused, ultimately, by the fact that uranium-238 was produced copiously in some supernovae several billion years ago.

Radiometric Dating

The fact that an atom's chemical reactions are largely independent of the nucleus leads to some very important ways of dating artifacts and rocks. The most familiar of these, the so-called carbon-14 dating method, is based on the fact that, in addition to normal, stable carbon-12, a certain amount of the radioactive element carbon-14 is always present in the environment. (Carbon-14 results from the collision of cosmic rays with nitrogen atoms in the upper atmosphere.) Because the chemistry of the two isotopes of carbon is identical, a certain amount of carbon-14 finds its way into all living tissues. When an organism dies, it stops taking in carbon-14, and its complement of these atoms starts to fall off as they decay. We know how much carbon-14 there is in the environment, so we know how much carbon-14 was in a piece of organic material when death occurred. Given that the half-life of carbon-14 is 5,730 years, we can work out how long it's been since that piece of material was taking in fresh carbon-14.

For example, if you find that a piece of wood contains only half the amount of carbon-14 it had when it was formed, you know that the tree from which it came died about 5,730 years ago. If the piece of material happens to be leather from a grave site or an elk shoulder blade used as a shovel, you have a pretty good idea of the age of the civilization that produced the artifact. This makes carbon dating an important tool in anthropology.

The same general technique can be used to date many rocks. The mineral's atomic structure tells us how much of a given isotope must have been present at the beginning, and measuring the amount left (or, equivalently, the number of decays that have occurred) will tell us how many half-lives it's been since the rock was formed.

One common technique for dating rocks involves the beta decay of potassium-40 to argon-40 (half-life: 1.3 billion years). Potassium is an essential element in many common minerals, while argon is a gas that is not incorporated into rocks when they form. If we crush and heat a sample of rock, each argon atom that we detect must be the result of one potassium decay after the rock formed. Knowing the number of decays that have occurred, and knowing how much potassium was in the rock originally, we can work out how long ago the rock formed. The potassium-argon technique was used to date four-billion-year-old moon rocks brought back by the Apollo astronauts, and it is employed routinely to date rocks on earth.

Radioactive Tracers

The independence of chemical reactions and nuclear reactions allows scientists to use radioactive tracers in fields as diverse as agriculture, geology, and medicine. The basic idea is simple: scientists introduce a sample containing a minute amount of a distinctive radioactive isotope into a system and follow its progress. The system's chemistry acts in its usual way on the sample, but the radioactive nuclei keep decaying, providing a "tag" that allows scientists to "see" the trace element as it moves, by chemical reactions, through the system.

Biologists commonly use tracers to follow the path of nutrients as they pass through the food chain. Physicians can watch iodine or thorium collect in the body and diagnose the presence of tumors. Earth scientists employ radioactive tracers to follow the paths of rainwater through groundwater reservoirs to lakes, streams, and wells. Oceanographers adopt the same technique to trace the direction and speeds of ocean currents. Anytime chemicals shift from one place to another, radioactive tracers can help document those movements.

Radiation Doses

We can measure two different characteristics of radiation: (1) how many particles a source gives off, and (2) how much potentially damaging energy is absorbed from that radiation. The first quantity, measured in units called curies, characterizes the source of radiation. The second number is the most important from the point of view of health risks, since it measures the effects of radiation on materials that it encounters. The familiar Geiger counter, which produces a click when it absorbs energy from passing radiation, was designed to measure this quantity.

One unit used to measure absorption is the rem ("radiation equivalent in man"). A dose of 750 rem is invariably fatal, and doses of hundreds of rem produce radiation sickness (but not necessarily death). Such high doses are usually encountered only in serious accidents or exposure to nuclear weapons.

At the other end of the scale, all living things on Earth are subjected to natural environmental radiation all the time. There is nothing sinister about this—the radiation was there long before the human discovery of nuclear physics and, for that matter, before there were humans at all. Coming from cosmic rays, radioactive isotopes in the air and ground, and even radioactive isotopes in our bodies, this dose typically amounts to 100 to 150 millirems (thousandths of a rem) per year for the average American. Add a slightly smaller annual dose from medical and dental X rays and you have a total average annual radiation dose in the neighborhood of 250 millirem. Whether this amount of radiation constitutes a health hazard is hotly debated by scientists at the present time, but there seems little prospect of a definitive answer any time soon.

FRONTIERS

Fusion Research

The prospect of harnessing fusion to supply our energy needs remains a dream of the scientific community. The central problem can be simply stated: how can you hold atoms together long enough, and raise their temperature high enough, to start a self-sustaining fusion reaction of the type that powers the sun? At present there are two general strategies to achieve this end. One, called magnetic confinement, involves holding atoms in intense, specially designed magnetic fields and heating them. The other, called inertial confinement, uses a small pellet of frozen hydrogen isotopes that is blasted with a laser beam. The blast causes the pellet to implode and heat up to the point of fusion.

Many steps lie ahead before fusion can become a viable energy source, and the press will surely tout each step as a breakthrough when it is completed. We must (1) demonstrate that we can produce a self-sustaining fusion reaction in the laboratory, (2) demonstrate that we can build a system that produces more energy than it uses, and (3) demonstrate that such a system can be economically competitive with other energy sources. The first of these steps may be completed by the end of this century.

The Structure of the Nucleus

The central problem of nuclear physics has always been this: given the properties of individual protons, neutrons, and the other particles inside the nucleus, can we predict the properties of nuclei made from them? It may surprise you that the answer is "no." Predicting the precise behavior of nuclei, even if there are only a few dozen neutrons and protons, is such a complex problem that it defies our best computers.

Today the focus of nuclear physics has shifted as scientists come to realize that all "elementary" particles that make up the nucleus are themselves made of smaller particles called quarks. The main question today is how this fact affects the structure of nuclei and, in particular, whether experiments can be devised that will shed light on the way quarks behave.

CHAPTER 9

PARTICLE PHYSICS

Deep in the heart of Texas, beneath the rolling range land of Ellis County south of Dallas, a mammoth construction project is about to begin. Billions of your tax dollars will be spent to excavate a 52-mile circular tunnel intended to house the largest physics laboratory in the world. The Superconducting SuperCollider (SSC), a huge accelerator, will be the largest high-tech construction project ever attempted. If completed, it will be the premiere machine on this particular frontier of science well into the twenty-first century. But its price tag ($8 billion or more—about $30 from every man, woman, and child in the country) has triggered a great deal of debate both inside and outside the scientific community.

Discovering the nature of matter is like peeling layers off an onion. Atoms are constructed from electrons and nuclei. Nuclei, though composed principally of neutrons and protons, are complex places where hundreds of different kinds of elementary particles whiz around, being created and absorbed each instant. And these particles themselves are not truly "elementary," but are made from things that are more elementary still.

Today we stand on the brink of discovering the ultimate "Theory of Everything" which may, after a two-millennium search, provide us with a detailed understanding of how our universe is built and ordered at the most fundamental level. At the foundation of that theory lies one key idea:

Everything is really made of quarks and leptons.

NUCLEI AND SUBNUCLEAR PARTICLES

Protons and neutrons are only two of the dozens of different particles found inside the nucleus of every atom. Starting in the 1950s, the frontier of physics has shifted from a study of the properties of nuclei to a study of the particles found inside them—a field called elementary particle physics, or high-energy physics. This remains the frontier today, although we know a great deal more about these particles than we did forty years ago.

The study of the particles inside the nucleus began in the 1930s when physicists studied cosmic rays. These high-speed particles, mainly protons, are produced in stars and rain down constantly on the earth from space. When cosmic rays hit a nucleus, two things can happen: (1) they can break the nucleus apart, spewing its constituent particles out where scientists can see them, and (2) some of the kinetic energy of the cosmic rays can be converted into mass, creating new particles.

To make a long story short, by the 1950s scientists had observed dozens of new kinds of particles coming out of collisions of cosmic rays with nuclei. All of these particles were unstable— they lived for only a short time and then decayed, like radioactive nuclei, into collections of other particles. Today scientists build huge machines called accelerators to produce beams of protons or electrons to use in place of cosmic rays. As a consequence, the roster of elementary particles has risen to several hundred.

Amid this proliferation of "elementary" particles, some general rules of organization become clear. There are basically two kinds of particles: those involved in structure (by far the majority) and those involved in forces. The proton, neutron, and electron, all basic building blocks of the atom, are examples of the first type of particles. You can think of them and other particles like them as the bricks of the universe—the things that, put together in different patterns, constitute everything.

The photon, the quantum of ordinary light, falls into the second class of particles. As we shall see shortly, the transfer of photons between charged objects creates the electromagnetic force that, among other things, holds electrons in their orbits. So the photon and its partners are the mortar of the universe, holding the bricks together. In the jargon of particle physics, they are called gauge particles.

Among the "bricks," there is a further important distinction. Some particles, like the proton and neutron, exist inside the nucleus and take part in the nuclear maelstrom. We call such particles hadrons, from the Greek for "strongly interacting ones." Other particles, like the electron, are not normally involved in the nucleus, but remain aloof from all the activity that goes on there. We call such particles leptons, from the Greek for "weakly interacting ones."

Physicists have one particularly strong prejudice about nature. They believe that, deep down, nature is simple. But as more and more subatomic "elementary" particles were discovered, things seemed to be getting very complex. Physicists counted at least four families of gauge particles (the photon, and other particles called the W and Z, the gluon, and the graviton) and six different leptons (the electron, and particles called the mu meson, and the tau meson, plus the three neutrinos associated with each of these), along with hundreds of different hadrons whizzing around inside the nucleus. But hundreds of particles can't all be elementary. If they were, it would violate the central belief of physicists that nature should be simple. As patterns of regularity among the hundreds of hadrons came to light in the late 1960s, scientists realized that hadrons are not themselves elementary, but are really collections of still more elementary things. We now call these more basic building blocks quarks (the term comes from a line in James Joyce's novel *Finnegan's Wake*).

At last particle physicists could breathe a collective sigh of relief. The idea of the quark *is* simple: there are only six different kinds of quarks, and different arrangements of them make all of the hundreds of hadrons, just as bricks can be combined to make an infinite variety of buildings. The proton and neutron, for example, are each made from three quarks.

The six kinds of quarks (or "flavors," to use the physicists' fanciful term) come in three pairs with the following names: up and down, strange and charm, bottom and top. Particles containing the first five flavors of quark have been seen in the laboratory, while a search is currently underway for particles containing the "top" quark.

So, in the end, everything is made from quarks and leptons. Quarks combine to make hadrons, and hadrons combine to make the nuclei of atoms. Electrons (which *are* leptons) attach themselves in orbit to make atoms, and atoms hook together to make all the infinite number of things we see about us. After a millennium and a half of searching for an answer to the question "What is matter?" many physicists believe we are close to the final answer, that we have peeled the last layer off the cosmic onion. Time will tell whether they are correct.

THE FOUR FORCES

We all have an intuitive notion of force as a push or pull. Modern physicists, however, see forces quite differently. The current view is that every force arises from the exchange of a particle. One way to picture this exchange is to imagine two skaters approaching each other, with one skater holding a bucket of water. As they pass, the skater with the bucket throws the water at the other. Both skaters will change direction—one because of the recoil of the throw, the other because of the impact. Newton's First Law says that any change of direction results from the action of a force. It is clear that the force in this instance is due to (a physicist would say "mediated by") the water exchanged between the two skaters. In exactly the same way, physicists picture all forces between elementary particles as being mediated by the exchange of gauge particles.

For example, when two electrons approach each other, physicists picture what happens as follows: One electron, like the skater who throws the water, emits a photon. The other electron, like the skater who gets hit, absorbs the photon. The result: the electrons recoil from each other, and we say a force acts between them. Even the force between large objects like a magnet and a nail is thought to be generated by floods of photons being exchanged between the two pieces of metal.

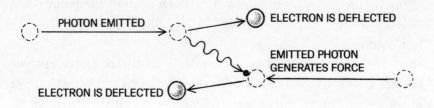

As two electrons approach each other, one emits a photon, which the other absorbs. This exchange results in a force between the two electrons — the electromagnetic force. This kind of illustration is called a Feynmann diagram after American physicist Richard Feynmann (1918–89).

There are only four forces that operate in nature. Two are familiar from everyday experience — gravity and electromagnetism. Two others operate at the level of the nucleus. The first, called the strong force, acts to hold the nucleus together against the electrical repulsion between the protons. The other, called the weak force, is responsible for interactions like the beta decay of nuclei and neutrons. Whenever anything happens in the world, it happens because one or more of these forces is acting. The four forces differ from one another because each involves the exchange of a different kind of gauge particle.

Every time you use your eyes or feel the sun's warmth on your skin you detect photons, the gauge particles associated with the electromagnetic force. Photons have neither mass nor electric charge, and they travel at light speed.

The strong force between quarks is mediated by a gauge particle appropriately named the gluon (it glues the quarks together). There are eight different kinds of gluons, all of them massless. None have been seen in the lab as yet.

The weak force is mediated by two related gauge particles called the W and Z. First observed at the CERN particle accelerator in Switzerland in 1983, these particles are over eighty times as massive as the proton. The production and study of the W and Z remains a major area of effort in current high-energy physics research.

No one has ever seen the gauge particle associated with the force of gravity, but physicists have a good idea of what it will be like. They believe that the ultimate theory of gravity requires a particle called the graviton. Physicists argue that the properties of the gravitational force require that the graviton will be massless and chargeless and travel at the speed of light, like the photon.

The Particle Zoo

There are so many kinds of elementary particles that sometimes it's hard to tell the players without a scorecard. We list below a few of the terms and particles that you might run across.

NEUTRINO: Neutrinos are massless, electrically neutral particles that are often emitted during radioactive decay. They are, for example, one of the products of neutron decay. The neutrino is a lepton—it does not take part in nuclear interactions. There are three different neutrinos—one paired with the electron, and one each with the mu and tau leptons (see below).

ANTIMATTER: For every particle, it is possible to produce an antiparticle. The antiparticle has the same mass as the particle, but is opposite in every other feature. For example, the antiparticle to the electron, the positron, has a positive electrical charge. When a particle meets its antiparticle, the two annihilate each other and all their mass is converted into energy.

MU AND TAU LEPTONS: These particles are just like the electron, but heavier. They do not participate in nuclear reactions. The mu (pronounced "mew") was discovered in cosmic ray debris in 1938, while the tau (pronounced "taw") was found in 1975 at the Stanford Linear Accelerator Center. There is a type of neutrino associated with each of these, just as there is an "ordinary" neutrino associated with the electron.

MESONS: Historically, a meson was any particle with a mass between that of an electron and a proton. Today, the definition is extended to include any particle whose decay products do not include a proton. There are mesons much heavier than the proton in the elementary particle zoo.

Accelerators

A large part of the cost of pursuing high-energy physics goes to build the machines called accelerators. As the name implies, these machines take particles (either electrons or protons) and accelerate them to speeds near the speed of light. These energetic particles are then directed against a target, where they collide with protons or nuclei. In the debris of these collisions, physicists search for the answers to their questions about the structure of matter.

From machines a few feet across (in the 1930s) accelerators have grown to mammoth structures many miles in diameter. In a typical accelerator, protons are injected into a large ring lined with magnets. The magnets exert a force that keeps the positively charged protons moving in a circular path, and each time they come to a certain point in the ring, their energy is boosted. In the modern generation of machines, the effective energy is increased by sending two groups of particles around the ring in opposite directions and arranging things so they can collide head-on.

There are three major accelerator centers you're likely to read about:

CERN: The European Center for Nuclear Research is run by a consortium of Western European nations. Located in Geneva, Switzerland, it has consistently enjoyed a position as one of the world's most important centers of high-energy physics. Its most recent addition is a ring nineteen miles in diameter in which electrons and positrons collide to create particles like the W and Z in abundance.

FERMILAB: The Fermi National Accelerator Laboratory, located outside of Chicago, is currently the world's most powerful accelerator. This machine's large ring, a mile across, accommodates circulating groups of protons and antiprotons that are made to collide head-on. Fermilab pioneered the successful use of magnets made from superconducting materials and remains the largest installation of superconducting wire in the world.

SLAC: The Stanford Linear Accelerator Center, located at Stanford University on the San Francisco peninsula, is one of the highest-energy electron accelerators in the world. The main working part is a two-mile-long tube down which electrons ride an electromagnetic wave like surfers on the ocean.

In a decade or so we may have to add the Superconducting SuperCollider to the list. By producing much higher collision energies, the SSC may probe even deeper into the fundamental structure of all matter, unlocking secrets now hidden from us.

UNIFIED FIELD THEORIES

A unified field theory is one in which two forces, seemingly very different from each other, are shown to be basically identical. In a sense both Newtonian gravity and Maxwell's equations represent unified field theories. The first showed that heavenly and earthly gravity were identical, the second that electricity and magnetism were really the same thing. Today the term is used to refer to new theories in which two or more of the four fundamental forces are seen to be the same.

You can visualize how seemingly different forces might be identical by thinking about the analogy of the ice skaters and the bucket of water. Suppose you had two sets of skaters—one set with a bucket of water frozen to ice, the other with a bucket of antifreeze—and suppose that the temperature in the rink was below freezing. The exchange that leads to the force might look very different—in one case it would involve a solid block of ice, in the other a fluid. We might argue that there were two different forces operating in the rink. If we raised the temperature, however, the ice would melt and we would see that the two forces were fundamentally the same—the similarity had been masked by the original low temperature.

In the same way, physicists argue that we have four forces today only because temperatures are low. If we allow particles to collide at very high speeds, the temperature at the point of collision will go up and we should see forces unify. The theories that predict how this unification will take place are the modern unified field theories.

The theories say that unification will take place in stages as the energy and temperature go up—two forces unifying, then a third joining those two, and finally the fourth coming in. The first unification, the one that brings together the weak and electromagnetic forces, has already been seen in the world's great accelerators. The theories that describe this unification did quite a good job of predicting things like the masses and production rates of the W and Z particles, so we have a great deal of confidence in them.

Physicists are paying a great deal of theoretical attention to the next unification—the one in which the strong and electroweak forces come together. The experimental evidence for the so-called Grand Unified Theories (or GUTs) is a bit murkier. Some predictions of these theories seem to be well verified, others not. The final unification, in which all four forces come together, remains the terra incognita into which theoretical physicists have only now started to venture. True to the tradition of fanciful nomenclature in particle physics, fully unified theories are often called TOEs ("Theories of Everything").

FRONTIERS

Quantum Gravity

The unification of gravity with the other three forces remains elusive, despite a decade of effort by the world's best theoretical physicists. The candidates for theories that might describe this unification go by names like "supersymmetry," "superstrings," and "quantum gravity." They all have the same goal: to describe the gravitational force in terms of the exchange of particles, just the way the other three forces are described.

Two Fundamental Problems

There are six kinds of quarks, six kinds of leptons. Both quarks and leptons seem to be found in groups of two each. Why is the universe arranged in this way?

Nature created the electron, then repeated the process again at higher masses for the mu and tau mesons. Why?

We don't expect answers to these questions soon, but they illustrate the sorts of problems that physicists will try to solve once the current generation of questions is answered. As we learn more about the fundamental structure of the universe, the overriding question of "Why this way and not some other way?" will occupy more and more of our attention.

The Top Quark

Right now, a major race is underway between CERN and Fermilab to be the first laboratory to discover the top quark, the only one of the six quarks for which evidence has not yet been found. The longer it takes for a discovery, the more likely it is that Fermilab, with its higher available energies, will win the race. When physicists discover the top quark, an important piece of the puzzle in our picture of the structure of matter will fall into place.

CHAPTER 10

ASTRONOMY

The sky is wondrous on a clear, cold, moonless night far from city lights. We marvel at the majestic sea of stars—thousands of stars, navigated by a half-dozen planets and the occasional brief meteor.

Scientists are no different in their sense of awe and wonder, and they turn to the stars in their search for answers to questions about the meaning of it all. To our unaided eyes all the stars look like brilliant points of light, some a little brighter and others fainter, some colored with a hint of red or blue. But when we focus our telescopes skyward we see many different kinds of stars. Some are hot and dense, burning their fuel at an incredible rate. Others are cool, consuming fuel much more slowly. We see stars in their infancy and stars growing old. And once in a great while we catch a glimpse of a star in its final cataclysmic hours, wracked by a massive fatal explosion. All this variety of stars tells a story:

Stars live and die like everything else.

The life of any star is a constant battle against the force of gravity, which tries to pull the star in on itself. Against this unremitting force, stars deploy a variety of countervailing strategies. Some of these strategies allow stars to stave off collapse temporarily and some strategies allow them to stave it off forever. But nothing can protect the largest stars from eventual collapse into a black hole, the ultimate victory of gravity over matter.

The Birth of Stars

All stars begin as diffuse clouds of dust in deep space. Somewhere in the cloud is a place where matter has gathered by chance more thickly than elsewhere, and the force of gravity exerted by the clump pulls in neighboring materials. This makes the clump more massive and increases its gravitational attraction, so even more material is pulled in. It's not hard to guess the outcome of this process—the cloud starts to collapse around the original concentration of matter.

As the contraction progresses, the pressure and temperature at the center increases. First, electrons are torn off their parent atoms, creating a plasma. Then, as the contraction continues, the nuclei in the plasma start moving faster and faster until, at last, nuclei approaching each other are moving so fast that they can overcome the electrical repulsion that exists between their protons. The nuclei come together and nuclear fusion begins—the nuclear fires ignite. Energy from fusion pours out from the core, setting up a pressure in the surrounding gas that balances the inward pull of gravity. When the energy reaches the outer layers, it moves off into space in the form of electromagnetic radiation and the stabilized cloud begins to shine. A star has been born.

The primary fuel for the fusion reaction is hydrogen. Two protons (the nuclei of hydrogen atoms) come together to form deuterium (an isotope of hydrogen consisting of one proton and one neutron) and some other particles. Subsequent collisions of the deuterium with other protons eventually produces helium-4, a nucleus consisting of two protons and two neutrons. In symbolic form, the nuclear reaction can be written as:

$$4 \text{ protons} \rightarrow \text{helium} + \text{energy} + \text{leftover particles}$$

As in nuclear reactions, the conversion of mass (in this case some of the mass of the four initial protons) supplies the energy.

While the star is contracting and stabilizing itself, some interesting events are taking place out in its periphery. The original cloud will, in general, have some small rotation. As the contraction starts, the rotation speed increases. The cloud is like an ice skater who, when she pulls her arms in while spinning, spins faster. If nothing counters it, contraction will increase the spin until the star is torn apart. There are two ways for the nascent star to avoid this fate: it can split into two, forming a double star system, or it can form planets. In both cases, the spin is transferred from the body of the star to the revolution of the stars or planets around each other. Most stars seem to take the double star route—at least two thirds of those you see in the sky are multiple star systems. The search for genuine planetary systems around other stars goes on, but astronomers can't yet prove that other systems like our own exist anywhere.

Stellar Lifetimes

The appetite of stars for hydrogen is truly prodigious. The sun, for example, consumes some 700 million tons each second, with about 5 million tons being converted into energy (primarily in the form of gamma rays). Yet so large is our star that it has burned its hydrogen at this rate for 4.6 billion years and will continue to do so for another five billion before running out of fuel.

How long will a star live? That, of course, depends on how much hydrogen it has and how fast it is consumed. Oddly enough, the larger a star is, the shorter its lifetime. The reason for this seeming paradox is simple: the bigger a star is, the greater is the gravitational force trying to make it collapse and the more hydrogen has to be burned to keep the star stable. The sun, a quite ordinary star, has enough fuel to keep gravity at bay for ten billion years, but a star thirty times as massive as the sun must burn its fuel in such a profligate way that it will shine for only a few million years. A star much smaller than the sun, on the other hand, will live for tens of billions of years—longer than the age of the universe. The star just pokes along, doling out miserly bits of energy into space as it husbands its hydrogen throughout a long and frugal life.

The Death of Stars

Profligate or miserly, every star must eventually burn up all of its hydrogen, filling the core with helium ash. When the hydrogen is gone, the outward force generated by the nuclear reactions disappears and gravity resumes its inevitable inward march. The inner parts of the star start to contract and warm up. For a star like the sun, the interior heating temporarily produces more energy as hydrogen burns just outside the core and the outer regions of the star are pushed farther outward, creating what astronomers call a red giant. Five billion years from now the body of the sun will extend out past the orbit of Venus, swallowing its two innermost planets and scorching the surface of the earth.

The core continues to contract even as the outer layers puff up, and soon the core becomes so hot that helium, the ash of the hydrogen fire, itself starts to fuse. In a series of reactions, three helium nuclei come together to form nuclei of carbon. Once the helium is consumed (a process that may take only a few minutes in a star like the sun), the collapse starts again in earnest. The bloated outer layers are blown off, while the inner region continues to contract. There is no more fuel to burn, so something else has to stop the collapse. That "something else," for the sun and stars like it, is related to the behavior of electrons. Electrons in the star cannot overlap—they need elbow room. Only so many can be crammed together in a given volume. When the core has collapsed down to the size of the earth, its electrons will have reached the point where they cannot be further compressed and the star will be stabilized forever, with gravity pushing in and the electrons pushing out. A star held up by pressure from its electrons is called a white dwarf. It generates no internal energy, having used up all its fuel, but continues to glow for a long time as it cools off.

Currently, theorists believe that stars with masses up to eight times that of the sun will end up as white dwarves. Made entirely of carbon nuclei, such a star is truly, as our childhood rhyme told us, "like a diamond in the sky."

If the star is very massive, however, its death is much more spectacular. It burns through its hydrogen quickly. Then, after a short collapse, it starts burning helium to make carbon. When the helium is exhausted and the inevitable collapse starts again, the temperatures at the center of the star get so high that even the carbon starts to fuse. This pattern continues, the ashes of each fire serving as fuel for the next as the star desperately tries to stave off the inevitable. In the final stages of nuclear burning, iron starts to be produced. Iron is the ultimate nuclear ash. It is impossible to get energy from iron by allowing it to fuse with another nucleus, and it is impossible to get energy from it by fission. As the star's core clogs up with iron, there is no way for the star to generate more energy. Again the collapse starts, but this time the force exerted by the electrons is not enough to overcome gravity. Electrons are forced inside the protons in the core, neutrons are produced, and the core shrinks quickly to a sphere of neutrons—a neutron star—about ten miles across. The force of gravity and the pressure of neutrons against each other balance and, providing the force of gravity is not too strong, the core stabilizes.

Supernovae and Their Consequences

When the core collapse occurs, the outer parts of the star find that the rug has been pulled out from under them. They start to fall inward, meet the rebounding neutron core and a flood of neutrinos created in nuclear reactions, and the star literally tears itself apart. For the space of about half an hour, shock waves crisscross the stellar carcass, creating temperatures in which all chemical elements up to uranium and plutonium are synthesized in a wild free-for-all, then blown into space. For a brief few days, the star can emit more energy than an entire galaxy. This event is a supernova—the most spectacular stellar cataclysm known. When the dust has cleared in a supernova, the end product may be a neutron star or a black hole—we don't know enough to predict which with certainty.

On February 23, 1987, a star exploded into a visible supernova in a neighboring region of our galaxy, giving astronomers a ringside seat during the spectacle and, not incidentally, giving them a chance to verify their theories of how stars live and die. The theories came through with flying colors.

When the fireworks of a supernova are over, all that's left of the original large star is a neutron core—a sphere of solid neutrons about ten miles across. The neutron star is usually rotating very fast, typically turning on its axis thirty to fifty times a second, because the collapse speeds up the original slow rotation of the star (remember the ice skater). The star's original magnetic field has also been concentrated by the collapse, and a field many trillion times that at the surface of the earth exists on the neutron star.

Electrons spiraling in toward the north and south magnetic poles of the rotating star give off energy, mostly in the form of radio waves. This radiation moves out into space in a narrow beam focused at the pole of the star. You can think of it as being something like the beam of a searchlight. As the beam sweeps by us, we see a pulse of radio waves, then darkness, then another pulse. When these signals were first seen in the radio sky, they were called "LGMs" (for "Little Green Men") because they looked like coded signals. Now we realize that they result from rotating neutron stars, which astronomers call pulsars. There are about 500 known pulsars in the sky, and probably many more waiting to be found.

If the star is very massive, then even the force exerted by the neutrons will not be enough to overcome gravity, and the collapse will continue down to a black hole. Black holes represent the ultimate triumph of gravity, the ultimate defeat of the star.

Our picture of the life of stars, then, is that early in the history of the universe large stars formed, lived their brief life, and became supernovae. In the last moments of their existence, these stars synthesized all the known chemical elements, then returned them to space. There, these elements in turn are incorporated into new generations of stars as the concentration of heavy elements increases throughout the universe.

All elements heavier than helium, including the iron in your blood and the calcium in your bones, are made in stars. We are, all of us, made of star stuff.

THE SOLAR SYSTEM

Our sun formed from a slowly rotating cloud of interstellar dust. Almost all of that cloud's debris was pulled into the proto-sun, but a tiny fraction of the mass concentrated instead into nine planets, plus a diverse collection of asteroids and moons, which adopted stable orbits around the sun. This collection of small, relatively cool objects is called the solar system.

The fact that the planets formed out of a contracting, heating, spinning ball of gas explains a number of the regularities we see when we look at them. For one thing, all of the planetary orbits lie in the plane of the sun's equator, and all of the planets move in the same direction around the sun. This systematic behavior occurred because the rotation of the collapsing cloud tended to throw material outward in the plane of the rotation, and it was in that plane that the planets eventually formed. You can think of the nascent solar system as something like a tennis ball (the sun) stuck in the middle of a large pancake, with the planets eventually forming in the latter. The process of formation itself is thought to be similar to the gravitational bunching described for the formation of a star.

Close to the sun, the temperatures in the cloud were high enough to vaporize substances like methane and ammonia. Particles streaming out from the sun blew these and other gases into space, leaving only solids behind to form the planets. This is why the inner solar system is populated by small, rocky planets. Farther out, however, methane, water, and ammonia were frozen solid and the original stock of hydrogen and helium was not greatly affected by the early sun. In the outer reaches of the solar system, then, we find the so-called gas giants—large planets formed primarily of frozen hydrogen, helium, methane, and ammonia.

Littered throughout the planetary systems are the remains of the building process—material that for one reason or another never got incorporated into larger bodies. The asteroid belt, between the orbits of Mars and Jupiter, contains the rocky remains of a planet that never fully formed, probably because of the gravitational influence of Jupiter. Far outside the orbit of Pluto is a swarm of slowly orbiting comets called the Oort cloud (after Dutch astronomer Jan Oort). Occasionally collisions or other disturbances in the Oort cloud send new comets into the inner solar system, where some (like Halley's comet) are captured by gravity into sedate, predictable orbits.

Connecting all these bodies is a thin, wispy web of magnetic field that starts deep inside the sun and extends outward to the galactic magnetic field. The magnetic fields of individual planets are embedded in the interplanetary field like lumps in gravy, and a steady stream of particles from the sun's surface, called the solar wind, moves out along the field lines.

A QUICK TOUR OF THE SOLAR SYSTEM

Terrestrial Planets

The planets Mercury, Venus, Earth, and Mars, together with Earth's moon, are usually designated the terrestrial (earth-like) planets. They are relatively small and rocky. Mercury and the moon are too small to hold gases to their surface, but the other three have atmospheres.

Venus is shrouded by clouds, but has been mapped by radar from orbit. Soviet spacecraft have landed on its surface, which is at a temperature of approximately 500°C (850°F). Of all the planets, Venus is the closest to the Earth in size.

The diameter of Mars is about half that of Earth. The planet has a thin atmosphere, mainly carbon dioxide, and the red color of its surface reflects the oxidized (rusted) iron in its rocks and soil. There is no evidence for the existence of life or liquid water on Mars, nor are there "canals," despite a mythology to the contrary. The American *Viking* spacecraft sent back spectacular images during its orbits and unmanned landings on Mars.

Jovian Planets

Jupiter, Saturn, Uranus, and Neptune are called the "Jovian" planets, after the Roman name for the god Jupiter. The largest

Jovian planet, Jupiter, has a mass more than three hundred times
that of Earth. These planets probably have a rocky core slightly
larger than the size of a terrestrial planet, but the core is buried
under thousands of miles of liquid and solid hydrogen, helium,
methane, water, and ammonia. All Jovian planets have multiple
moons and ring systems, with the rings of Saturn being the most
spectacular and best known. They are far from the sun, and
therefore cold. Some of their moons are virtually planets in their
own right, being larger than Mercury. All of the Jovian planets
have now been observed close up by the *Pioneer* and *Voyager* spacecraft.

Pluto

Pluto is normally the farthest planet from the sun, though its
elliptical orbit occasionally takes it inside the path of Neptune for
part of its year. Pluto is small and rocky with one large moon.
Because of its un-Jovian appearance, some scientists think Pluto
may be a large captured comet rather than a true planet.

GALAXIES

Stars are not scattered at random throughout the universe; all
stars are gathered into clumps called galaxies. Our own sun, for
example, is one of a group of about 100 billion stars called the
Milky Way galaxy. About 120,000 light-years across, the Milky
Way, like about three fourths of all galaxies, is a flattened rotating
disk with bright spiral arms. On a good night, your unaided eye
can see about 2,500 stars, all of them in the Milky Way. You can
also see a few fuzzy patches of light called nebulae ("clouds").
Seemingly unimportant in the grand celestial display, they are
actually other galaxies, with billions of stars, planets, and possibly life like our own.

The discovery that there were other "island universes" besides
the Milky Way was made by the American astronomer Edwin
Hubble in 1923, using the then new 100-inch telescope on Mount
Wilson, near Los Angeles. Until this telescope was built, astron-

The M 81 spiral galaxy, which contains billions of stars, is one of the closest galaxies to our own Milky Way. Viewed from 4 million light-years away, our home cluster of stars would look similar to this. PHOTO COURTESY OF THE CARNEGIE INSTITUTION OF WASHINGTON.

omers, like someone peering at fine print without glasses, had been trying to understand nebulae without much success. The advent of the Mount Wilson telescope changed all that. Hubble could pick out types of stars that astronomers used to establish distances within the Milky Way. Using measurements of these stars, Hubble showed that the Andromeda nebula is some 2 million light-years away—far outside the confines of the Milky Way. Because of his work, we now realize that our galaxy is only one among billions, each with billions of stars, in the universe.

Most galaxies, like the Milky Way, are relatively sedate, homey places where the slow process of the stellar life cycle goes on quietly. A small number, however, seem to house a kind of violence unknown in our own peaceful neighborhood. Cataclysmic explosions rip through the cores of the galaxies, spewing huge jets of material hundreds of thousands of light-years into space. These active galaxies typically emit large amounts of energy as radio waves, and hence shine brightly in the radio sky.

The most interesting of the active galaxies are the quasars (the name comes from a contraction of "quasi-stellar radio source"). A quasar may easily emit more energy in a second than the sun has in its entire lifetime. The thousands of quasars known in the sky tend to lie at great distances from Earth—in fact, the most distant objects known are quasars. One current theory is that quasars are an early, violent stage in the evolution of galaxies. According to that theory, light from quasars has been traveling toward us for billions of years, in some cases from the very beginnings of the universe. The Milky Way may once have been a quasar and may appear to be one now to astronomers at the other end of the universe.

Telescopes

Hubble's work illustrates an important point about astronomy. Our knowledge of the universe is intimately tied to our ability to build large (and expensive) telescopes to detect and record the radiation that comes to us from deep space. As a rule of thumb, a state-of-the-art ground-based telescope costs about as much as a major highway interchange. That was true when Mount Wilson was built in the early twentieth century, when its more illustrious 200-inch partner was built on Mount Palomar in the 1930s, and, given current rates of inflation, will be true when the next generation of telescopes is finished.

There are two major areas of telescope building going on today: ground-based and satellite. Palomar represents the pinnacle of building large "light buckets" from single blocks of glass. Today, telescopes use modern fast electronics to achieve the same result much more simply. The Keck telescope, built by the California Institute of Technology at Mauna Kea in Hawaii, typifies this new departure in design. It is made of many small mirrors—in fact, its working surface looks like a plateful of potato chips. Each of these small mirrors is individually controlled by a computer that constantly adjusts the total array to keep everything in focus. In this way, a series of small mirrors can take the place of a single large block of glass, producing a telescope that is, in terms of its light-gathering ability, more powerful than Palomar.

In the same way, radio telescopes no longer are being built as gigantic single dishes—overgrown versions of home TV antennae—but as series of individually controlled receivers whose positions are synchronized by a central computer. The largest such telescope is the VLA (Very Large Array) located in the desert near Socorro, New Mexico.

Only visible light and radio waves can penetrate through earth's atmosphere to ground-based telescopes. To monitor the other parts of the electromagnetic spectrum, we must lift receivers above the obscuring air. Instruments located in satellite observatories have greatly expanded our understanding of the universe. In the future, permanent satellite telescopes are planned for infrared and X-ray astronomy. These observatories go by the name of SIRTF (Satellite InfraRed Telescope Facility) and AXAF (Advanced X-ray Astronomy Facility).

The best-known orbiting observatory is the Hubble Space Telescope (HST), which was launched in April 1990. Primarily designed to detect visible and ultraviolet radiation, when it is working properly the HST will allow scientists to look at distant objects with unprecedented resolution. It will not, however, see farther into space than the best ground-based telescopes.

FRONTIERS

Search for New Planets

Do other stars have planetary systems? Planets do not give off visible light—we see our neighbors through reflected sunlight only—but they do emit infrared radiation. One continuing frontier of astronomy will be the search for planets around other stars. To date, some evidence exists for large partners of some small nearby stars, but these might be small double-star partners rather than planets.

SETI

The search for extraterrestrial intelligence, or even extraterrestrial life, catches the imagination but does not have high priority among astronomers. We know from the space program that it is extremely unlikely that life will be found in our solar system. There are programs to monitor nearby stars (which may or may not have planets) for radio messages sent by extraterrestrials, but they tend to be rather small-scale operations. Some scientists argue that the human race is probably alone in the galaxy, because it is extremely unlikely that all the conditions necessary to produce intelligent life will be present nearby in the galaxy. For the record, however, we feel that these searches should be carried out. If we find someone out there, the implications will be staggering. If we don't, the implications will be even more staggering.

CHAPTER 11

THE COSMOS

Where did the universe come from? Where is it going? How is it put together? How did it get to be the way it is?

These are Big Questions. Like others of their ilk, they are easy to ask and very hard to answer. We want answers to them for deep philosophical reasons—reasons that have very little to do with the immediate applications of technology. No one is going to get rich from discovering the structure of the universe (unless, of course, she decides to write a book about it).

The branch of science devoted to the Big Questions is called cosmology. Modern thought in this area derives from the fact that:

The universe was born at a specific time in the past, and it has been expanding ever since.

THE EXPANSION OF THE UNIVERSE

Edwin Hubble established the existence of other galaxies, but this was not the most important result of his work. When he looked at the light coming from those galaxies, he found that it was shifted toward the red. That is, its wavelength was longer than that of the light emitted from corresponding atoms in laboratories. Furthermore, he found that the farther away the galaxy was, the greater was the shift in its light. Hubble attributed this "red shift" to the Doppler effect.

You experience the Doppler effect every time a speeding car passes you as you stand on the sidewalk. Sound waves have a regular pulse or frequency, and your ear interprets that frequency as a pitch. If a noisy object like a horn or a racing motor moves toward you, then you hear more pulses per second because the source moves a short distance toward you between sound-wave crests. Your ear hears a higher pitch. Once past, however, the receding vehicle imparts fewer pulses per second as the source moves away from you between crests. As a result, the pitch sounds lower. Listen for that characteristic change in pitch, from high to low, the next time a loud vehicle whizzes by.

Light can experience a Doppler shift just like sound. Light emitted by a star that is speeding away from us appears to be at a lower frequency (shifted toward the red end of the spectrum) than that emitted in a laboratory. Hubble saw this red shift and concluded that almost all galaxies are rushing away from us and the universe as a whole is expanding. Observations with modern instruments verify that this so-called Hubble expansion exists throughout the observable universe. This fact is central to our present ideas about the universe.

Imagine a piece of rising bread dough with raisins scattered throughout it. Each raisin represents a galaxy, the dough the space that separates galaxies. If you were standing on one raisin, you would see the neighboring raisin receding from you because the dough between you and it is expanding. A raisin twice as far away would be receding twice as fast because there is twice as much dough between you and it. The farther away the raisin, the faster it would be moving. This is exactly the sort of behavior that Hubble observed for galaxies.

The bread dough analogy illustrates several important points about the Hubble expansion. First, there is no significance to the fact that the earth seems to be the center of the universal expansion. In the bread dough, you see the same thing: no matter which raisin you stand on, it appears that you are standing still and everyone is moving away from you. Thus, everyone sees himself as the center of the universal expansion, and the fact that we see everything moving away from the earth in no way makes us special.

Second, the movement of the raisins is not like the explosion of an artillery shell. The raisins do not move through the dough, but are carried along by the general expansion. In the same way, the galaxies do not fly apart through space, but are carried along as space itself expands.

Third, the raisins do not themselves expand—only the space between them. In the same way, the solar system and the Milky Way galaxy are not expanding, even though distant galaxies are receding from us.

Finally, if you ask where in the dough the expansion started, you can only answer "everywhere," since a bit of dough at any point in the bread now was at the center when the expansion started. In the words of the fifteenth-century philosopher Nicholas of Cusa, "The universe has its center everywhere and its edge nowhere."

THE BIG BANG

The Hubble expansion has one remarkable and inescapable consequence: it requires that the universe had a beginning. If you think of "running the film backward" on the current expansion, you find that fifteen to twenty billion years ago the universe was a single geometrical point. The current expansion must have started at that time. The initial event, as well as the general model in which the universe started expanding from a highly condensed beginning, is called the Big Bang. It represents our best guess as to the origin and evolution of the universe.

Universal Freezings

When the universe was younger, it was denser—more compressed—than it is now. When matter and energy are concentrated in a small volume, temperature inevitably is higher. Consequently, when the universe was younger it was hotter. Tracing backward in time, we can recognize six crucial events—we like to call them "freezings"—where the fabric of the universe changed in a fundamental way, much as water changes when it freezes into ice. Understanding these freezings is the main task of modern cosmology.

The most recent freezing occurred when the universe was about 500,000 years old (i.e., about 14,999,500,000 years ago). After the first 500,000 years electrons and nuclei formed permanent attachments in the form of atoms, but prior to that time if an electron fell into orbit around an atom, it would be knocked off by a collision with another speeding particle. Before 500,000 years, matter existed as loose electrons and nuclei—the state of matter we have called plasma.

Moving backward in time, the next freezing occurred at about three minutes, when nuclei formed. Before this time, there were only elementary particles in the universe, and if a proton and neutron came together to form a nucleus, they would be torn apart by subsequent particle collisions. After three minutes nuclei could remain stable (although for reasons discussed below, only nuclei up to helium and lithium were formed in the Big Bang—everything else was made later in stars).

From three minutes back to about ten millionths of a second the universe was a seething mass of elementary particles—protons, neutrons, electrons, and all the rest of the particle zoo. At ten millionths of a second the universe had cooled off enough so that quarks could come together to form the elementary particles. Before this time there were only leptons and quarks, afterward there were leptons and the whole sea of elementary particles that live inside the nucleus.

The First Ten-billionth of a Second

Ever since the universe was ten millionths of a second old, the great freezings involved changes in the fundamental state of matter. Before that time there were three more freezings, each involving forces rather than matter. When quarks "froze" to form the elementary particles, the forces acting in the universe were pretty much as we see them today. There were four distinct forces—the strong, electromagnetic, weak, and gravitational. But earlier in the history of the universe, when things were hotter, some or all of these forces must have been unified. One by one, they come together as we move backward in time until finally, at the very beginning, there is only a single, all-encompassing force.

The timetable for the unification of the forces as we theorize them today is as follows:

1/10,000,000,000 second: the weak and electromagnetic forces unify into one force, called electroweak. The temperatures of the universe at that time can be reproduced on earth at accelerator laboratories. Thus, we can have some confidence in our understanding of the universe from the time it was one ten-billionth of a second old to the present, because we can test in the laboratory our theories of what happened then.

1/1,000,000,000,000,000,000,000,000,000,000,000 (or 10^{-33}) second: the strong force unifies with the electroweak, leaving only gravity as odd man out. During this freezing, two other important events occurred: the entire universe expanded rapidly from something smaller than an elementary particle to something the size of a grapefruit (a process known as "inflation"), and antimatter started to disappear, annihilating itself with matter to produce radiation. The Grand Unified Theories that describe this freezing make predictions about laboratory experiments, and therefore can be tested. The results of these tests are, thus far, inconclusive.

1/10,000,000,000,000,000,000,000,000,000,000,000,000,-000,000 (or 10^{-43}) second: known as the Planck time (after one of the founders of quantum mechanics), this marks the ultimate unification. Until the universe was this old, all four forces were unified, and things were as beautiful and simple and elegant as they could be. Particles of matter in its most fundamental form interacted through the medium of a single unified force. It's all been downhill since then.

EVIDENCE FOR THE BIG BANG

The Cosmic Microwave Background

No matter which direction you look from Earth, microwave radiation from deep space is raining down. This radiation, discovered in 1964, constitutes the first great verification of the Big Bang. The reason is this: every object gives off radiation, and the type it gives off depends on its temperature. Your body, for example, gives off infrared radiation because you are at a temperature of 98.6 degrees Fahrenheit. If the universe started from a hot beginning and has been expanding and cooling off ever since, it would now be at a temperature of about three degrees C above absolute zero. A body at this temperature emits microwaves, and it is this radiation that we detect on Earth.

The First Elements

The formation of nuclei at the three-minute mark lasted for only a short time before the Hubble expansion spread matter too thinly for nuclear reactions to proceed. During this short burst of nucleus building, there was time to create appreciable amounts of deuterium (one proton, one neutron), various forms of helium (two protons, one or two neutrons), a little lithium (three protons, four neutrons), and nothing else. We can calculate the amounts of deuterium, helium, and lithium nuclei produced by estimating the temperature of the three-minute-old universe and the known rates at which nuclear reactions proceed.

When we look at the universe, we find that the predictions for these primordial abundances are remarkably well borne out (once we have corrected for the elements later created in stars). These predictions are exact and unforgiving, since a few percent deviation in helium abundances from predicted values would rule out the standard Big Bang model. The fact that our predictions have proved to be so accurate probably constitutes the best available evidence in favor of the Big Bang.

THE STRUCTURE OF THE UNIVERSE

Just as stars are collected into galaxies, galaxies themselves are collected into structures known as groups, clusters, and super-clusters. The Milky Way and the Andromeda galaxy, for example, form the gravitational anchors of what astronomers call the Local Group, which is made up of these two plus a gaggle of twenty or so smaller galaxies. The Local Group, in turn, lies at the edge of the Local Supercluster, which consists of about 100,000 galaxies.

Almost all the known mass of the universe is gathered into su-perclusters, which can be pictured as groups and clusters strung together like beads on a string. Between the superclusters lie regions known as voids—desert-like volumes where almost no stars shine. These voids, which can be millions of light-years across, were totally unknown until the early 1980s, when modern data-analysis techniques allowed astronomers to pick out the empty areas despite the fact that light from galaxies behind them shines through and reaches the earth.

The best way to visualize our present picture of the universe is to imagine cutting through a mound of soapsuds. You'd see a series of empty bubbles, each surrounded by soap film. Replace the film by superclusters and the bubbles by voids and you have the universe.

The Final Frontier

The main task that this generation of cosmologists faces is to find the laws that governed the first fraction of a second of the Big Bang, subject to the rather formidable constraints that those laws must produce a universe in which matter is clustered into gal-axies, and galaxies are clustered into superclusters separated by voids, but the cosmic microwave background is the same no mat-ter which direction we look. Finding such a theory is no trivial task, and a lot of bright people have tried and failed. It seems as if the more we learn about the universe's structure, the harder it gets to make all the pieces mesh. This leads some scientists to hope that when we finally do find a theory that works, it will also be the *only* theory that does so—a true Theory of Everything.

FRONTIERS

How Did Galaxies Form?

Until atoms formed, the matter in the universe interacted readily by radiation. If matter had started to come together into galaxies before the 500,000-year mark, the forces exerted by radiation would have blown those concentrations apart. Once atoms formed, light could move through the matter easily and the universe became transparent. Only then could the gathering of matter into galaxies have begun. But soon after 500,000 years, the Hubble expansion would have spread matter out so thinly that galaxies could not have formed at all. This is the central problem of galaxy formation, compounded today by the fact that any theory that provides a way around this Catch-22 must also explain why galaxies themselves tend to form superclusters, separated by vast intergalactic voids.

The best guess of cosmologists is that invisible "dark matter" may constitute an abundant, but as yet undiscovered, form of mass in the universe. This hypothetical material does not scatter radiation, and therefore could have clumped together much earlier than ordinary matter. The idea is that the dark matter was already grouped into the superclusters when atoms formed, and thereafter attracted ordinary matter into a preexisting structure.

By observing the movement of hydrogen atoms far from the spiral arms of galaxies, astronomers have concluded that there is a lot more matter in galaxies than the stuff that shines. Some researchers believe that this newly discovered material makes up at least 90 percent of the mass in the universe. Discovering the location of dark matter and how much there is stands as a frontier of observational astronomy, while trying to imagine the nature of that dark matter remains a challenge for theorists. One thing seems clear, however. Whatever the dark matter is, it's not like anything we've made in our laboratories up to now.

Where Did the Universe Come From?

Although the question of where the universe came from occurs instantly to most people when they think about the Big Bang, it is not something that gets a lot of attention from cosmologists. Scientists don't yet feel we have the tools to attack the problem intelligently, although a few intrepid souls have started to blaze the trail for us.

One point to note is that there is no problem in principle with creating matter from a vacuum. Matter is just another form of energy, and can be produced if the energy input is balanced by something else. For the universe that "something else" could be negative energy in the gravitational field. If this were the case, creating the universe would be like digging a hole—you'd have a pile of dirt (visible matter) balanced by a hole (the gravitational field). The process is miraculous only if you ignore the hole and insist that that matter appeared "from nothing."

Is the Universe Flat?

If there is enough mass in the universe, the Hubble expansion will eventually stop, perhaps to be reversed. The theories that predict the inflation at 10^{-33} second also predict that the total amount of mass in the universe exactly equals what is needed to stop the expansion. In the jargon of astronomers, the universe is expected to be flat. Counting dark matter, we can account for about 30 percent of this expected mass. Where, if it exists, is the rest?

CHAPTER 12

RELATIVITY

Our chapter on relativity is quite different from the others in this book. You have day-to-day experience with motion and forces, matter and energy. Chemicals, the earth, and living organisms are tangible things. But relativity requires an abstract, even philosophical, approach to science. Albert Einstein discovered relativity not by performing experiments, but by thinking about the way nature must be. You can follow Einstein's ideas by doing the same simple thought experiments.

Relativity is a fascinating subject because it gives you a whole new way of looking at the universe. Still, a lot of scientists, ourselves included, would place relativity pretty far down the list of things you need to know to be scientifically literate. But Einstein and his remarkable theory are part of our cultural heritage— scientific folklore, if you will. Relativity is fun—and it's great for impressing people at parties.

Let's get one thing straight right from the start. It is absolutely untrue that there are only a dozen people in the world who understand the theory of relativity. This statement may have been true in 1920 (although, to tell the truth, we doubt that it held even then). Today the basics of relativity are routinely taught to college freshmen in "Physics for Poets" courses, and hundreds of graduate students in physics and astronomy learn the full mathematical rigors of relativity every year.

Relativity is not technically difficult. Indeed, the most basic statement of relativity may surprise you by its simplicity:

Every observer sees the same laws of nature.

Relativity has one central precept: there is no "correct" place from which to view the universe—no "God's-eye" view of things. Every observer, whether sitting in a rocking chair on earth or traveling near the speed of light in deep space, sees the same laws of nature. The only laws that could possibly govern the universe, according to the theory of relativity, are the same everywhere you look.

This idea seems almost too simple, yet many of the theory's results violate our intuition about the way the world ought to behave. Relativity requires us to face up to the fact that the world doesn't always behave the way we expect it to, and many people find it unsettling that nature is so indifferent to our sense of the rightness of things. If you can get used to the idea that the universe is what it is, regardless of what we think it should be, then you'll have no problem with relativity.

REFERENCE FRAMES

When you sit in a chair at home, you observe the world from a *frame of reference* that is firmly attached to the solid earth. If you ride by in a car or a plane or a spaceship, on the other hand, you observe the world from a frame of reference that is moving with respect to the earth. In either case you are an *observer* in the relativistic sense: in either reference frame you could set up a physics lab and perform experiments. In either frame of reference you could describe physical phenomena and deduce the laws of nature.

No matter what your reference frame, you can think of yourself as being stationary while every other observer is moving. This view may not seem obvious—when you drive you probably don't think of your car as stationary while the countryside whizzes past. Most of us are accustomed to thinking of the earth as the "right" frame of reference, and we unconsciously put ourselves into the earthbound frame of reference whether we are moving or not. But have you ever, while sitting in an airplane being pushed back from a gate or a bus backing out of a station, thought, just for an instant, that the plane or bus next to you was moving forward? In that moment, before your conscious mind took over and reimposed its prejudice, you were a true relativistic observer. Your frame of reference was your own fixed center of the universe, and everything was moving around you.

Different observers give different descriptions of the same event. If someone riding in a car drops this book, the book falls straight down as far as the passenger is concerned. But watching this event from the side of the highway, you would see the book fall in an arc—the car's motion carrying it along some distance during the time it takes to fall. You and the person in the car would give different descriptions of the fall.

But now suppose that you and the person in the car each equip yourselves with a physics laboratory and each of you determines the laws that govern the fall of objects in your own frame of reference. When you compared your results, you would find that they were identical—you would both have come up with Newton's Laws of Motion. In other words, observers in different frames of reference give different descriptions of specific events but identical descriptions of the laws that govern those events. This is the central idea of the theory of relativity. If we assume that it is a general truth, then we can derive its consequences and test them experimentally. In the end, we shall see that the predictions that follow from this principle meet the test of experiment, and this is why scientists accept the theory.

The principle of relativity can be stated simply as "Every observer sees the same laws of nature," but in practice, it is easier to break this principle into two parts, based on how the observers and their frames of reference are moving. The easier part we call "special relativity," and it deals with the special case of reference frames that do not accelerate. In the Newtonian sense, special relativity is concerned only with observers in uniform motion, with no forces to alter their motion.

On the other hand, "general relativity" applies to all frames of reference, whether they accelerate or not. General relativity contains special relativity as a special case, but is itself a lot harder to deal with mathematically. We'll treat special and general relativity separately, tackling the simpler theory first.

One more point should be made about the place of relativity in modern scientific research. Although the theory of relativity has the aura of being in the forefront of modern science, in fact it has been around since 1905 and is regarded by physicists as a familiar and well verified part of their world. Except for attempts to merge general relativity and quantum mechanics, very little research on relativity per se is done these days. The results of the theory are simply incorporated into work on other subjects.

SPECIAL RELATIVITY

If the laws of nature must be the same for all observers moving at constant speed, then all such observers must agree with Maxwell's description of the laws of electricity and magnetism. Since the speed of light is a constant built into Maxwell's equations, it follows that all observers must measure the same value for the speed of light. If they didn't, different observers would find different sets of Maxwell equations.

This conclusion, in and of itself, already violates our intuition. Think about a simple example: you stand on a railroad car moving 50 miles per hour and throw a baseball forward at 50 miles per hour. Naturally someone on the ground will see the baseball moving at 100 miles per hour—the speed of the train plus the speed of the baseball. But if instead of throwing a baseball you shine a flashlight, the principle of relativity says that someone on the ground must see the light moving at a speed of 186,000 miles per second, not 186,000 miles per second plus 50 miles per hour.

This sort of basic contradiction between the way we think the universe ought to behave and the idea that the laws of nature have universal validity first led Albert Einstein to think about the theory of relativity. In the late nineteenth century, there were three ways in which the problem could have been resolved:

1) Maxwell's equations could have been wrong; or,
2) the principle of relativity could have been wrong; or,
3) our intuitive ideas about space and time could have been wrong.

The last possibility arises because to calculate speed, we have to divide the distance traveled by the time it takes for the travel to occur. In our intuitive thinking about the problem of the flashlight, for example, we assumed that a clock on the ground and a clock on the railroad car would both run at the same speed. In reality, this might or might not be true—you never know until you actually make measurements.

During the 1920s, serious theories proposed modifications to Maxwell's equations to make the speed of light depend on the motion of the source. When these modifications were tested (for example, by measuring light emitted by a double-star system both when the star is coming toward us and going away), they failed. In fact, we can say that both Maxwell and Einstein have been abundantly supported by the data. That leaves only the third possibility, that there is something wrong about our intuitive notions about the way different observers see things like clocks and yardsticks.

According to Einstein, he came to the realization that moving clocks might not run at the same rate as stationary ones while riding in a streetcar in Bern. Looking at a clock on a tower, he realized that if the streetcar moved away from the clock at the speed of light, it would look to him as if the clock had stopped. Einstein would, in effect, be riding outward from the clock perched on a single crest of a light wave. His pocket watch, on the other hand, would be moving with him and hence would continue to tick along in its usual manner. Thus, he reasoned, it is at least worth considering the possibility that our usual assumption that time is the same for all observers is simply wrong when things move at speeds near the speed of light.

We instinctively think something is wrong with the idea that light travels at exactly the same speed whether from a moving or stationary flashlight. We base our prejudice on a lifetime's experience with moving objects. But how much of this experience was garnered while moving at speeds close to the speed of light? None of us has ever moved even close to 186,000 miles per second, so, strictly speaking, we have no experience whatsoever about how light or baseballs should behave at such speeds. The only thing actually violated by the example of the relativistic flashlight is our untutored expectation that nature should be the same at high velocities as it is at low ones. But this is only an assumption and, like other assumptions, it must be tested against experiment before we can accept it.

Let's explore some of the remarkable consequences of the principle of relativity with an open mind and see how the predictions of the theory stack up against experiment.

Time Dilation

A clock ticks off equal bits of time. Anything that "ticks" can be used to measure time's passage. Think about a clock made from a flashing strobe light, a mirror, and an instrument that records the arrival of a light beam. Each "tick" of the clock consists of a light flash, the transit of light to the mirror and back, and the click (or whatever) of the instrument when the light returns. If the arrival of the light in the instrument triggers the next flash of the bulb, the clock will "tick" regularly. You could imagine adjusting the distance to the mirror so that the "ticks" of the light clock were synchronized with the ticks of any other kind of clock—the pendulum of a grandfather's clock, the vibrations in your quartz wristwatch, and so on. Despite its strange appearance, the "light clock" is a perfectly ordinary clock.

Imagine two light clocks, each held perpendicular to the ground, one next to you and the other going by in a car moving at a constant velocity. We arrange things so that both bulbs flash as they pass each other. The light in the stationary clock travels up to its mirror and back. Meanwhile, the light in the moving clock moves upward as the entire clock travels along to the right.

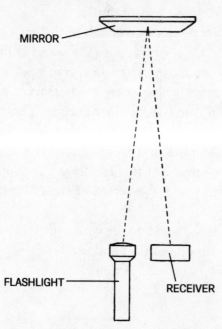

MIRROR

FLASHLIGHT

RECEIVER

The "light clock" is constructed from a flashing light, a mirror, and a receiver. Each "tick" of the clock is the time it takes for light to make the round trip.

As a result, the light in the moving clock *as seen by an observer on the ground* must travel in a sawtooth pattern as shown.

The principle of relativity says the speed of light must be the same in all frames of reference. Light in the moving clock has to travel a longer distance so it must take longer to reach its destination than light in the stationary clock, which only has to travel up and down. The ground-based observer will see both lights flash, then he will hear his own clock tick, and only later will he hear the moving clock tick. This pattern will be repeated with each click, and the moving clock will fall farther and farther behind its stationary counterpart. If the speed of light is the same for all observers, it follows that *moving clocks run slower.* This effect is known as time dilation.

Many people's first response to this argument is that it is built on an illusion—that the moving clock isn't "really" running

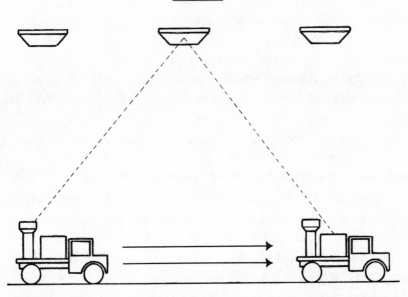

Albert Einstein realized that a moving light clock appears to be running at different speeds for different observers. The faster a clock moves relative to the observer, the farther its light must travel, but the speed of light is constant. (The mirror moves along with the truck.) Einstein concluded that as a clock approaches the speed of light, time appears to move more slowly.

slower. As teachers, we have learned to recognize the use of "really" as a tipoff to a Newtonian frame of mind, and to use this objection to bring our students face to face with the real core of the theory of relativity. For the fact of the matter is that when someone says that the moving clock doesn't "really" slow down, what he or she means is that the clock appears normal *to an observer moving with it.* In the jargon of the physicist, the clock appears normal in its "proper" frame of reference.

The assumption hidden in the objection is that somehow the proper frame is "right" and other frames are "wrong," and that only the proper frame should be consulted if you want to know what the clock "really" is doing. But the central thesis of relativity is that there are no "right" frames of reference—no privileged positions from which events ought to be viewed. Every observer—every frame of reference—has an equal right to be heard when descriptions of physical events are given.

What is even more disturbing than the counterintuitive notion of time dilation is that the effect actually exists in nature. There are many proofs of this statement, but the most dramatic experiment was performed by scientists at the University of Michigan. They strapped extremely accurate atomic clocks into first-class seats on aircraft making round-the-world flights and, after the journey was completed, compared the readings on those clocks to readings on clocks that had been left on the ground. As expected, the moving clocks had slowed down slightly.

So time dilation, as well as being easy to derive from the principle of relativity, is also supported by experiment. As difficult as it may seem to square with our intuition, clocks in moving frames of reference run more slowly than stationary ones. In our normal experience this slowdown is much too small to measure except with the most precise instruments: a clock that had been moving at 60 miles per hour since the beginning of the universe would have lost only one second by now.

If the velocity of the moving clock is small compared to the speed of light, the width of the sawtooth will be small and the distance traveled by the two light beams will be almost the same. It is only when the sawtooth spreads out (i.e., when the velocity of the clock approaches that of light) that appreciable differences appear. You don't have to give up your intuition about clocks for everyday experience quite yet, but if humans ever develop space ships that travel near light speed, time dilation may wreak havoc with future genealogists. Slowly aging space travelers could return from a voyage younger than their earthbound children!

You can test your understanding of time dilation by convincing yourself that an observer traveling next to the moving clock will think that the clock on the ground has slowed down.

Other Predictions of Special Relativity

Einstein called the kind of exercise we just went through for the moving clock a "*gedanken* experiment" (from the German *denken* — to think). It's a technique that allows you to grasp the essential behavior of things like clocks, even though performing that particular experiment might be technically difficult. Other "*gedanken* experiments" allowed Einstein to draw startling conclusions from his special theory of relativity:

1) Moving yardsticks are shorter than stationary ones.

When an object moves, it contracts — physically shrinks — along the direction of motion. Thus, a baseball moving at or near the speed of light will look flattened, like a cookie viewed edge-on.

2) Moving objects are more massive.

The faster an object moves, the greater its mass becomes and the harder it is to deflect it from its course. As its velocity approaches the speed of light, the mass of any object approaches infinity. This result leads to the common misconception that nothing can move faster than the speed of light. Relativity doesn't say that at all — it just says that nothing now moving at less than the speed of light can be accelerated to and past that speed. There's still room for Warp Drive!

3) $E = mc^2$.

The most famous outcome of the theory of relativity is the equivalence of mass and energy. This simple equation has been elevated to the level of folklore, perhaps the only equation of physics to enjoy that status. It says, in effect, that mass is just one more form of energy. Mass can disappear provided an equivalent amount of energy in another form takes its place. More strikingly, if there is a lot of energy available (for example, in the collision between two particles), some of that energy can be converted into mass, and a new particle can be created where none existed before. The new particle isn't created "out of nothing," but from energy taken from another source.

Because the speed of light, c, is such a large number, the conversion of a little bit of mass can produce a lot of energy. By the same token, it requires a great deal of energy to produce even a small particle. A block of cement small enough to fit under your kitchen table could run the entire United States for over a year if it were converted completely to energy.

Experimental Confirmation

Scientists have been able to verify all of these predictions of special relativity. For example, physicists routinely use particle accelerators to take bunches of protons and electrons up to speeds near that of light. The speeding particles are kept in a designated track by large magnets, and the force exerted by the magnets has to be adjusted to take account of the fact that the particles' masses increase. Every time one of these machines operates it confirms predictions of the theory of relativity.

By the same token, accelerators work by manipulating long, strung-out bunches of particles. As the particles accelerate, these bunches shorten, and the machine is adjusted to take this effect into account. Thus, the fact that accelerators work is also evidence for the prediction of length contraction.

Finally, about 20 percent of all electrical power in the United States is produced by nuclear reactors. Reactors work because nuclear reactions convert small amounts of mass into large amounts of energy, in keeping with Einstein's famous formula. Thus, the equivalence of matter and energy is confirmed every day by commercial power companies.

Some Philosophical Remarks About Relativity

In many ways, the philosophical consequences of the theory of relativity are as important as the practical results that flow from it. It was the first of the modern theories that revolutionized the old, mechanical, Newtonian view of the world. Relativity substituted equal observers for the classic approach by which all laws were referred to a single, correct "God's-eye" frame of reference. But relativity did not consign Newton to the garbage dump of history. It simply extended our knowledge into ultra-fast domains that Newton never investigated. When we apply the equations of relativity to the modest speeds where Newtonian mechanics has worked in the past, relativistic equations reduce to the same ones that Newton first wrote down three centuries ago. So Einstein didn't really replace Newton; he encompassed and expanded Newton's work.

Finally, and perhaps most important, the theory of relativity is not simply a statement that "everything is relative," even though it is usually expressed that way in casual party chitchat. What is "relative" in relativity are descriptions of specific events. But the crucial aspects of knowledge, the laws of nature, are most emphatically not relative. Every observer in the universe, regardless of his or her state of motion, must obey the same physical laws. Thermodynamics, Maxwell's equations, and quantum mechanics apply to every observer, and do not vary from one to another.

GENERAL RELATIVITY

Imagine yourself in a windowless spaceship accelerating at exactly one "G," the equivalent of earth's gravity. Could you tell if you were in space or on earth? The answer is no. If you drop a ball in the spaceship, as far as you are concerned the ball falls to the floor. Looking at this event from a stationary frame outside the spaceship, we would say that the floor had accelerated upward and hit the stationary ball. But inside the sealed-off ship the ball appears to fall, just as it would on the earth. No experiment you could do would tell you whether you were in an accelerating spaceship or stationary on the earth. Thus, acceleration and gravity must be equivalent at some deep level, and what we call gravity must be an effect of our frame of reference. This equivalence is the central thesis of Einstein's theory.

The central idea of general relativity is that someone in an accelerating frame of reference (such as a rocket ship) experiences exactly the same effects as those normally associated with the force of gravity. Einstein saw a connection between changes in motion (what Newton would have called the action of a force) and the geometry of reference frames. The result of his thinking, published in 1916, was the general theory of relativity—the theory that still stands as our most complete theory of gravitation.

The long wait between the special and general theories was due primarily to the complexity of the mathematics Einstein had to develop to express his ideas. The fact that theoretical physicists are now trying to replace the theory with one more attuned to the notions of quantum mechanics in no way diminishes relativity's impact on science. We will first present a simple way of visualizing general relativity (minus the complex mathematics), then discuss some tests that support it.

Imagine taking a sheet of tough plastic and stretching it tightly over a large, rigid frame. Imagine further that you had painted a rectangular grid on the flat surface. If you carefully rolled a lightweight ball bearing along one of the grid lines, it would continue to roll along the straight line. The only way you could get it to deviate from the grid line would be to exert a force on it (e.g., by blowing on the ball bearing or by bringing up a magnet). This picture represents the classical Newtonian way of looking at all forces, including gravity. Objects move in a straight line unless a force causes them to do otherwise.

General relativity approaches the problem of motion in a completely different way. Imagine placing a heavy lead ball on the sheet of stretched plastic we've just discussed. The lead sphere will weigh the plastic down, distorting and warping it. If you now rolled the ball bearing across the plastic, it would follow a path that carried it nearer the lead sphere than it would otherwise have gone. Newton would say that an attractive force (such as gravity) existed between the two, but Einstein would interpret the same phenomenon differently. He would say that the presence of the lead weight warped the space in its vicinity, and that this warping caused a change in the ball bearing's motion. For Einstein, there were no forces in the Newtonian sense, only changes in the geometry of space.

The relativistic interpretation of the solar system, then, is that the sun warps the space around it and the planets move around in this space like marbles rolling around the inside of a bowl. In fact, if you use relativity to calculate what happens to the original grid of straight lines when you drop a mass onto it, you find that the original lines are deformed into closed elliptical curves—precisely the paths followed by planets. A good way to keep things straight is to remember that:

For Newton, motion is along curved lines in a flat space.
BUT
For Einstein, motion is along straight lines in a curved space.

Einstein believed that not just gravity, but all forces would ultimately be explained in this geometrical way. In fact, he spent the last half of his life in a futile search for a unified theory of forces. Progress toward this goal came about only after yet another way of describing forces—through the exchange of elementary particles—had been developed. Thus, general relativity remains a magnificent but isolated chapter in the sciences; Einstein's theory encompassed and superseded Newtonian gravity, and will itself soon be encompassed and superseded by a theory of quantum gravity.

Tests of General Relativity

Unlike special relativity, general relativity is not buttressed by lots of experimental evidence. The reasons for this lack of evidence are partly theoretical and partly technical. Like special relativity, general relativity encompasses Newton's physics. For normal, everyday phenomena, general relativity gives predictions that are virtually the same as those of Newtonian gravity. Thus, unless we are able to make extremely precise measurements, the two cannot be distinguished in a laboratory setting. It is only in the region of very large masses or very short distances that the warping of space becomes so pronounced that the two theories differ significantly, and such conditions are not available to experimenters.

There are only three classic tests of general relativity. These are (1) the precise shape of planetary orbits, (2) the bending of light near the sun's rim, and (3) the gravitational red shift.

Because planetary orbits are elliptical there is a point where the planet comes closest to the sun. We call this point the perihelion (from the Greek for "near the sun"). In a simple Newtonian situation, the perihelion would be at the same point in space through all time—the orbit would not shift. In fact, many forces act to push the perihelion of a planet a little farther along each time it goes around. The gravitational effects of the other planets, particularly Jupiter, are the most important. But before Einstein published his theory, the measured perihelion advance of the planet Mercury exceeded the predicted value by about 43 seconds of arc per century. General relativity predicted that the (very small) warping of space by the mass of the sun would produce exactly this much advance in the perihelion. This retrodiction was rightly taken to be a great triumph for the theory.

Today scientists use radar ranging to make extremely accurate determinations of the orbital positions of the planets, and the perihelion shifts for Venus, Earth, and Mars have been measured and found, like that of Mercury, to be precisely as predicted by general relativity. This is probably the most stringent test of the theory available at this time.

The best-known test of general relativity involves the bending of light rays that pass close to the rim of the sun. The measurement of this predicted effect in a 1919 eclipse catapulted Einstein to a position of international prominence. Today, we perform this test with radio waves instead of light, and the sources of the radiation are quasars, not stars. Radio waves can be detected any time (they are not blotted out by the sun), so scientists can conduct this test under normal conditions, rather than having to wait for an eclipse. The measurements agree with the predictions of relativity to better than one percent, another remarkable verification of the theory.

Finally, relativity predicts that as a photon climbs upward in a gravitational field (as it would in leaving the surface of the earth), some energy is drained from it by the uphill motion. In this the photon is no different from a baseball, which slows down as it gains height. Since the photon must continue to move at the speed of light, however, we see its loss of energy as a lengthening of the wavelength of the light—a shift toward the red. Thus, a flashlight beam seen from an airplane will be slightly redder than that same beam seen from the ground. Like the above two predictions, this too has been borne out by experiment.

For over half a century, there have been only three tests of general relativity, but each has seemed to supply supporting evidence. As we shall see, however, this lack of experimental tests is changing, and we can expect new results in the near future. Whether they confirm the theory or not remains to be seen, of course, but we would be willing to bet that when all the dust has cleared, Einstein will remain top dog.

Black Holes

The most spectacular prediction of general relativity is the existence of black holes. To understand the odd behavior of a black hole, go back to the analogy of the stretched plastic sheet and the lead weight. Imagine that we had a way of adding more and more mass to the lead sphere without increasing its size. As the ball got heavier and heavier, the distortion of the plastic would get bigger and bigger. Eventually the plastic might deform to the extent that the weight would neck off and separate itself from the rest of the surface. The distorting plastic sheet might even close up completely, wrapping the lead ball out of sight for good and leaving a reformed plastic sheet in its wake.

In just the same way, relativity predicts that when a large enough mass is concentrated in a small enough volume, it distorts the space around it so severely that a part of space wraps itself up and leaves the rest of normal space behind. A mass that has done this is said to have formed a black hole. You can think of the black hole as an object so massive and so dense that nothing, not even light, can muster enough energy to escape from its surface. Once something falls in, it can never get out. A material that absorbs all light that falls on it is black, which is how this particular beast received its ominous name.

Theorists think about two kinds of black holes. One is the end state of the implosion that accompanies the death of very massive stars. These objects, a mile or more across, are supposed to litter the galaxy like junk cars on a country road. There may also be "quantum" black holes, hypothetical objects smaller than elementary particles, that some theorists think exist inside the nucleus.

There is at present no conclusive evidence for either kind of black hole, although there are a number of candidates for the astronomical variety. The problem is that, by definition, you can't "see" a black hole by looking at electromagnetic radiation. Any radiation that strikes the hole is absorbed, never to reappear. All you can do is look for indirect effects caused by the gravitational effects of a black hole. You can, for example, search for distortions in light rays coming near the hole, or you might see X rays emitted by accelerated matter as it falls into it. You can look for double-star systems that behave as if one of the members were a black hole.

We agree with astronomers who suspect that several well-studied astronomical systems contain black holes. It is probably only a matter of time until their identity is proved to everyone's satisfaction. Quantum black holes remain much more problematical, and theorists must work out a lot more of their expected properties before a serious experimental effort can be made to find them.

FRONTIERS

Relativity, despite its flashy aura, is actually a staid and settled part of physics. It has become a tool that cosmologists and particle physicists use to understand the origins of the universe and the basic structure of matter, rather than a field of study in and of itself. In this respect, at least, it resembles Newtonian science.

The only area that might be called a frontier is the experimental testing of general relativity. This field is active right now because advances in electronics have finally given experimenters the ability to measure many of the extremely small differences that are supposed to exist between general relativity and Newtonian physics.

Perhaps the most striking of the upcoming tests of general relativity will be launched in a satellite experiment sometime in the mid-1990s. Twenty years in development and planning, this experiment consists of carefully machined quartz spheres set into rotation. Relativity predicts that the presence of the rotating earth will cause the axis of rotation of the spheres to wobble a little, and this small deviation from normality will be measured. This experiment requires amazing precision. The quartz spheres must be so perfectly round that, were they blown up to the size of the earth, their highest "mountains" would be no more than a foot high!

Over the next decade, a series of these sorts of high-precision experiments will be attempted, and physicists may at last have accurate verifications of the predictions of general relativity. If any experiments fail to confirm the theory, it will be big news. If they don't, we'll have the luxury of knowing that the theory still holds. Either way, the outcome will be welcome.

CHAPTER 13

THE RESTLESS EARTH

Imagine driving along the Nimitz Freeway in Oakland, California, late on the afternoon of October 17, 1989. The setting sun illuminates San Francisco Bay. Nature seems in perfect balance. Even the rush-hour traffic is unusually light, most people having left work early to watch the third game of the World Series.

Suddenly, beneath the small town of Loma Prieta, fifty miles to the south, rock strained by centuries of slow movement finally snaps, releasing its energy like a coiled spring. More powerful than a nuclear bomb, shock waves travel through the earth, damaging buildings and bridges thoughout the Bay area. The freeway supports fracture and sections of the roadway collapse around you, trapping and killing dozens of people in their cars. Just as suddenly, all grows silent. Then sirens begin to whine as frantic rescue workers speed across the ravaged city.

Earthquakes and volcanoes offer dramatic testimony that our planet is not at rest. For centuries humans viewed these destructive natural phenomena as purely chance events, governed by the whims of gods and unpredictable in their savagery. But earthquakes and eruptions do not occur randomly. You can live a long life as a resident of New York City, where earthquakes are very rare, and never feel the slightest tremor, much less worry about a volcano erupting in Central Park. But if you live along the California coast for just a few years the chances are that you will feel the earth shake, and if you reside on the big island of Hawaii for just a few months you will probably witness the eruption of a nearby volcano.

Even if scientists can't predict exactly when a "big one" will hit, they can tell you why earthquakes and volcanoes occur in certain parts of the globe. Earthquakes, volcanoes, mineral deposits, and even the oceans and the continents themselves are the surface manifestations of tremendous forces at work deep within the earth. Only within the past thirty years have scientists finally begun to understand the forces that operate to change the face of our planet—sometimes, as in San Francisco, in violent ways. The central fact that governs their new insights into the workings of our planet can be summed up as follows:

The surface of the earth is constantly changing and no feature on earth is permanent.

The forces that drive this constant change are generated deep inside the earth, where nuclei of radioactive elements constantly decay. The energy of decay is converted to heat, and this heat slowly seeps up toward the surface. Over hundreds of millions of years, rock warmed by the radioactive decay rises slowly to the surface, cools, and then sinks to be warmed again. The earth beneath us, viewed over a long time, is really no different from a pot of boiling water on your stove.

Encasing this roiling interior, called the mantle, is a layer of rock usually less than thirty miles thick, floating on the churning material like an oil slick on boiling water. In response to this churning, the thin outer skin of our planet breaks up, moves around, and reassembles in unceasing movement. And on top of this thin layer of restless rocks, like scum riding on an oil slick, are the continents—the part of the globe that we, in our arrogance, call "solid earth."

The "boiling" of the interior rocks causes the earth's continents to float around, collide, tear apart, and link together again. In the process, ocean basins open and close, mountain ranges are thrown up and weathered down and the surface constantly changes. Alone among the planets of the solar system, the earth

is restless. It is the only planet still in the process of forming itself—still being born.

The roiling surface of the earth is very like the surface of a
boiling pot of water, except that instead of a fluid moving around,
as in the pot, it is the solid rocks of the earth's interior that
"boil." To be sure, this boiling process is very slow.

Rocks rarely travel faster than about an inch a year, but give a
convecting rock a million years and it can travel miles. In a hundred million years it can move the length of a continent.

The heat that drives this motion of the mantle comes from two
sources: radioactive decay of materials in the mantle rocks, and
heat left over from the formation of the earth. Scientists sometimes argue over how much heat comes from each source, with
most favoring radioactive decay as the main source. From the
point of view of the mantle, though, the source of heat is irrelevant. However the heat got there, it has to be moved to the surface
by convection and then radiated into space.

PLATE TECTONICS

Earth scientists of all persuasions have embraced an elegant
planetary model that identifies a simple underlying cause for the
earth's violent moods. This model, discovered in the 1960s and
dubbed "plate tectonics," describes the interaction between
"plates," which are thin, brittle slabs of crust, and the vast mobile mantle that lies beneath the crust and makes up four fifths of
the solid earth. "Tectonics" comes from the Greek root "to
build," so plate tectonics refers to a theory of how the surface of
the earth is built from plates.

Plates and Continents

Our current picture of the earth's outer layers is simple. The
mantle is completely covered by a crust of basalt, a brittle, dense,
dark volcanic rock. If you've ever enjoyed Hawaii's black sand
beaches or passed by the rusty brown cliffs and columns of the
Hudson River Palisades in New York, you've seen basalt. As the

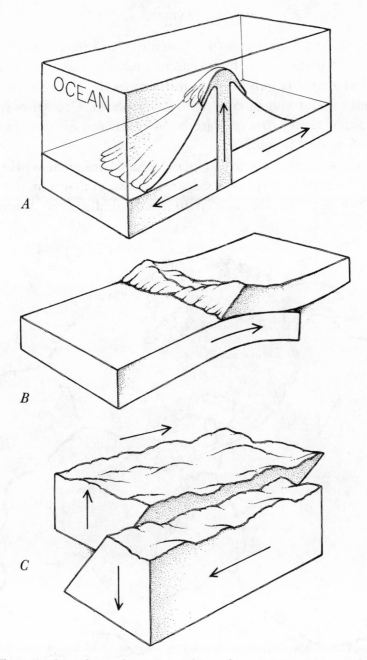

The great plates that make up the earth's surface can interact in several ways. (A) They can diverge along a volcanic ridge where new crust is constantly being formed; (B) they can converge so that one plate must be subducted beneath the other; or (C) they can scrape against each other to form a fault zone, which may produce violent earthquakes. In each case, the plate motions result from movement of the convecting mantle below.

mantle rolls and boils, the earth's basaltic cover breaks into many thin, brittle plates, each hundreds or even thousands of miles across but generally only three to thirty miles thick. These plates then move about and interact with other plates. Geologists recognize three main types of plate boundaries: divergent, convergent, and neutral boundaries.

New crust forms at *divergent* plate boundaries, places where mantle convection moves plates apart and brings new material to the surface. A great mountain chain in the middle of the Atlantic

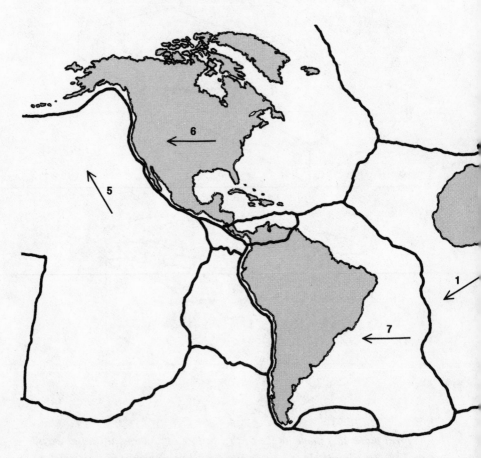

A number of major plates, along with many smaller pieces, form the earth's surface. Arrows at the plate boundaries of this map indicate the relative direction of plate motions.

Ocean (the Mid-Atlantic Ridge) marks one such zone of upward mantle convection. The island of Iceland lies along this ridge and consists entirely of volcanic rock. Other diverging boundaries occur beneath continents. If you visit the Great Rift Valley in East Africa you can actually stand on a place where the mighty African continent is just beginning to be torn apart.

If new plate materials are forming at diverging boundaries, then old material must be destroyed somewhere else, since the earth itself is not getting any larger. This destruction occurs at

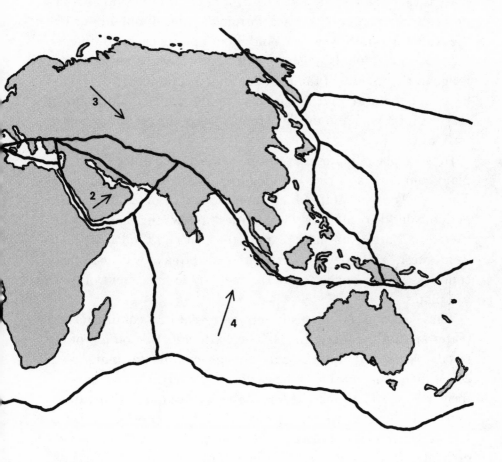

convergent boundaries where the plates are pushed together, driving one beneath the other, or "subducting" it. The subducted plate goes back into the mantle, where it can melt and mix with mantle rocks to join the reservoir of material ready to start the whole cycle again.

The surface manifestations of a subduction zone depend on whether the colliding plates are carrying continents or not. If neither has a continental passenger, the result is a deep ocean trench like the Marianas Trench near the Philippines. If only one plate is carrying a continent, the continent crumples up as it approaches the region of contact, forming chains of mountains like the Andes in South America. And if both plates carry continents, then those continents crash together and form a high mountain range in the middle of the new, combined continent. The Himalayas are the scar that resulted when the Indian Subcontinent crashed into Asia, and the Alps remind us of the event that joined Italy to Europe.

In some places plates may scrape against each other at a *neutral* plate boundary, scouring out long, destructive earthquake zones. The San Andreas fault, which runs the length of California and periodically disrupts West Coast cities, follows the boundary between the North American plate to the east and the Pacific plate to the west. The 1989 San Francisco earthquake was just one of countless shocks triggered by the inexorable movement of these two plates.

These examples of plates in motion reveal that continents and plates are not the same things. The continents ride on top of the basalt plates, but there is only enough continental material to cover about one quarter of the earth's basalt crust. Consequently, the earth is (and always has been) about three quarters ocean. At present the earth has only six continents, but we recognize at least a dozen major plates and there are probably a host of poorly defined smaller plates as well. The South American, Australian, and Antarctic continents lie within the boundaries of much larger single plates. The North American plate contains almost all of the continent of North America and about half of the Atlantic Ocean. The Eurasian and African plates, on the other hand, are composites of several continental masses. Continental blocks that were once separate, such as India and Asia, have fused. Ancient continents, fancifully named Gondwanaland, Laurasia, and Pangaea by geologists, have been pulled apart.

Plate tectonics reveal to us that nothing on the surface of the earth—no river or valley, no ocean or plain, not even the tallest mountain on the largest continent—is permanent.

DISCOVERY OF THE PLATES

Scientists of the 1960s were not the first to suspect that continents move. In 1912 the German meteorologist Alfred Wegener constructed a theory he called "continental drift" to explain, in part, why the west coasts of Africa and Europe so closely match the east coasts of the Americas. Wegener explained the fit by theorizing that the continents were once joined and somehow drifted apart to adopt their present positions. This explanation precisely fits plate tectonics today, but Wegener's continental drift hypothesis did not really anticipate the modern theory. His model just happened to have one thing—the motion of continents—in common with the modern view.

The acceptance of plate tectonics by the scientific community in the 1960s is one of the great events in the history of science. Despite the fact that almost all geologists considered moving continents to be the rankest sort of heresy, when the data began to accumulate they willingly abandoned the beliefs of a lifetime in deference to the new information. Every scientist must be willing to change his or her mind when the data require it.

The most striking evidence for plate tectonics came from an unlikely source: measurements of rock magnetism on the ocean floor. When molten rock comes to the surface, as it does wherever plate boundaries diverge, it usually contains small grains of iron minerals. These grains act like tiny compasses, lining up to point to the north pole. When the rocks solidify, the grains are locked in, and the rock "remembers" where the north pole was when the cooling occurred.

The magnetic field of the earth has reversed its polarity frequently over geological time. Your compass needle now points north, but a million years ago it might have pointed south. Iron minerals formed from molten rock a million years ago thus have the opposite orientation of more recently formed minerals.

In the 1960s ocean scientists discovered distinctive patterns of magnetic stripes in rocks on both sides of what we would now call diverging boundaries. As new rock rises from the mantle and fills the space left as plates diverge, the iron grains point north and become locked in that position. As times goes by and the older rock moves aside to make way for new material, the earth's magnetic field eventually undergoes one of its reversals and iron grains in those rocks point in an opposite direction from those of their older neighbors. Repeated reversals of the field produce the striped pattern, which can *only* occur in an environment where new crust is being created continuously *and* the magnetic field is undergoing sporadic reversals.

Plate tectonics suggests that plates will move a few inches per year. Until fairly recently, the idea that continents move was buttressed only by indirect evidence such as that provided by rock magnetism. No one had actually measured continental motion. In 1985 new confirmation for plate tectonics arrived from an unexpected source: extragalactic astronomy. In that year astronomers announced the result of measurements of the radiation from distant quasars. They measured the difference in arrival times for radio waves from the quasars at three observatories: one in Massachusetts, one in Germany, and one in Sweden. From these measurements, they obtained a very precise number for the distances between those observatories. In just a few years, this distance has grown by more than a foot as the separation between Europe and North America slowly increases, confirming that the continents do, indeed, move.

Why Plate Tectonics Is Important to Science

Before the 1960s, scientists who studied the earth tended to work in isolation from one another. The oceanographers measured currents and temperatures, but they never talked to the paleontologists who studied fossils, and neither group spoke to the geophysicists who probed the earth's deep interior. The various groups seemed to have little in common.

The advent of plate tectonics has changed all that. The model has supplied a common language, a common paradigm, and common ground of concern to all scientists who study the earth. Oceanographers now know that what goes on under the earth's crust affects ocean basins, paleontologists routinely use the evidence available in fossils to track the wandering of the continents across the globe, and geophysicists understand the earth's interior as a dynamic convecting system that drives the restless crust. Scientists now see planet Earth as a single integrated whole, rather than a series of isolated systems that have nothing to do with one another.

With the discovery of plate tectonics, a lot of seemingly random geological data began to make sense. Seismologists had known that most earthquakes strike in broad circular belts around the world, but they didn't know why. We now know that those belts coincide with scraping and colliding plates. Volcanoes are most common in long chains of young mountains; we now see that those mountain chains correspond to plate boundaries. Mining geologists found that the largest ore deposits, precipitated in the hot mineralized waters of volcanic districts, often occur above subducting plates; new deposits have been discovered as a result. And for the first time geologists and paleontologists can explain why distinctive ancient rock formations and fossil deposits match up across vast oceans. The simple idea of plate tectonics illuminates and unifies much of today's earth science research.

Earthquakes

Efforts to predict earthquakes reveal the strengths, as well as the limitations, of plate tectonics. We now know why major earthquakes shake the Los Angeles and San Francisco areas from time to time. Two massive plates are inexorably grinding past each other, and in the process California is being ripped apart. But knowing why doesn't necessarily tell us when. At the present rate of movement—a few inches per year—a major quake should hit every fifty to one hundred years. But specific events are almost impossible to pinpoint. Sometimes, "swarms" of little quakes precede major shocks, but it is not practical to evacuate Los Angeles or San Francisco every time a few small earthquakes are registered.

We do know that over time, as the two plates whose boundary is the San Andreas fault move in opposite directions, stress builds up deep in the earth. The process is like winding up a spring: eventually the rock fails and the energy is released. We can measure the amount of strain in rocks near the surface, and thereby guess where large earthquakes will most likely occur. But at the moment all we can say with certainty is that another "big one" will happen sooner or later. The 1989 San Francisco earthquake is regarded by some geologists as little more than a warmup for the large release of energy they expect to happen sometime in the next century—an impression reflected by their naming it "the pretty big one." At the moment, that's the best our science can do.

The media usually describe the strength of an earthquake by the Richter scale, introduced in the 1930s by the California seismologist Charles R. Richter. Richter assigned a value of 0 to the weakest rumble he could measure with his equipment. Each increment of 1 in the scale means ten times more earthquake energy, so a magnitude 4 earthquake (a noticeable event) is 10,000 times stronger than one of magnitude 0. The Richter scale is completely open-ended—any number is possible. Today's sensitive seismometers record earth twitches much weaker than 0 (they are given negative numbers). The Loma Prieta earthquake registered about 7 on the scale, while the strongest catastrophic earthquakes measured have been close to 9.

Why There Are No Mountain Chains on Mars

Earth is unique among the sun's rocky planets. Mercury, Venus, Mars, and our moon are unchanging worlds. Why should our

globe be different? Why don't our neighbors also possess continents that ride on mobile plates?

The critical factor is size. The other worlds are small enough
that all the heat generated by radioactivity inside leaks out by
conduction as fast as it is produced. You experience a similar
effect every time you eat hot food: a potful of soup can stay hot
for hours, and a bowl stays hot for several minutes, but a spoonful
looses its heat in a matter of seconds. Mars, Mercury, and the
moon, all mere spoonfuls of earth-like material, have long since
frozen to inert balls. Any new heat generated in their interiors
quickly flows to the surface and radiates away. They have no
plates to collide, no great earthquake fault zones or chains of
mountains. Venus, which is only slightly smaller than the earth,
may once have had its own sluggish version of plate tectonics and
may even have active volcanism today. But Venus also proved too
small, and its interior apparently no longer convects. The earth,
because it is slightly larger and traps its internally generated
heat, continues to roll and boil. Given enough time, it too must
cool and stop changing, but that won't happen for billions of
years.

A WINDOW INTO THE SOLID EARTH

The deepest mine descends only about two miles; the deepest
borehole penetrates less than ten miles. Scientists are usually
wary of placing limits on what humans may eventually achieve,
but at present we cannot conceive (even in our wildest fantasies)
of a way to journey to the center of the earth. Given such physical
limitations, how can anyone possibly know what's in the earth's
deep interior? One group of scientists, seismologists, use sound
waves to unlock earth's hidden secrets.

Seismology is a global-scale variation of sonar. Sonar measures
the time it takes for a sound wave (the ubiquitous "ping" sound
of submarine movies) to travel to an object (the ocean bottom or
another submarine), bounce off, and return. Seismology is almost
the same thing. Instead of a "ping," seismologists use dynamite
or earthquakes to generate a loud enough sound wave to travel

through the earth. Anything softer would be lost in the noise of landslides, construction equipment, and interstate traffic. The time it takes for sound to travel through the planet depends on the kind of rock through which it travels. By measuring many waves traveling along many different paths from an earthquake or explosion, the seismologists gradually build up a picture of the earth's deep interior.

The result of seismic explorations is an understanding of the interior of the earth as a series of concentric layers. The innermost layer, called the core, is about 2,600 miles in radius and made primarily of heavy metals like nickel and iron. The inner part of the core is solid, but the outer layer is a sea of liquid metal. Temperatures in the core may reach 7,000 C—enough to vaporize any known material at the earth's surface.

Surrounding the core and reaching to within a few miles of the surface is the mantle. Made of lighter materials, it is the mantle rocks that move slowly in response to the heat in the earth's interior and whose motion, ultimately, results in tectonic activities at the surface.

Finally, the outer surface of the earth, or crust, contains the mountains and valleys, oceans and plains, that make up our familiar surroundings. The crust contains the lightest materials in the earth—those that floated to the top when the planet was molten.

Most seismologists work for oil companies or mining concerns and study small-scale geological features, usually only a mile or so across. They wear rugged clothes and sturdy boots and travel from site to site with a drilling rig and a truckful of detectors called seismometers. The exploration team drills a hole, packs it with explosives, sets off the charge, and records the seismic echos in the hope of finding telltale rock structures that might indicate nearby deposits of valuable minerals or oil.

Government and academic seismologists often study the earth on a much larger scale. They have established hundreds of permanent listening stations around the world. These stations play a vital role by monitoring the location and severity of earthquakes. By comparing the arrival time, duration, and strength of seismic waves at many different stations, scientists can deduce the exact location and force of each quake. Civic planners depend on that information to predict zones of future earthquakes and thus guide development.

It may seem that earthquake seismologists spend most of their lives waiting for something bad to happen, but these earth scientists also play a key role in preserving peace, since they provide the technical basis for verifying nuclear test ban treaties. An underground explosion, which pushes rock out in all directions, has a different seismic "signature" than natural earth movements, during which rocks slide against each other. No large-scale nuclear test can escape the notice of the global array of seismometers. By computer analysis of the signals, scientists can determine the place and size of any underground blast, even at a distance of thousands of miles.

FRONTIERS
Searching for Buried Treasure

Plate tectonics directly affects how much you pay at the gas pump, for a knowledge of the positions of ancient plates and continents can lead us to untapped natural resources. One of the major challenges facing today's oil and mining geologists is to unravel the tectonic history of the earth. Geologists and geophysicists can discover the locations of ancient plates and continents, oceans and mountain ranges by integrating many types of studies—fossil distributions, rock magnetism, field mapping, and seismology.

Oil fields formed eons ago from thick accumulations of organic materials in tropical or temperate zones. Early in this century, before anyone conceived of the notion of wandering continents, no one would have predicted the discovery of major oil reserves in the Arctic regions of Alaska, but with our new picture of moving plates and continents it is obvious that once-tropical lands could end up literally at the ends of the earth. The search for fossil fuels has expanded accordingly.

Plate tectonics has also changed the way we look for metal mines. Many metal deposits lie near ancient plate boundaries, where hot volcanic mineral waters concentrated ore, so modern prospectors study the history of the earth's wandering plates. Rich mines of gold in China, copper in Chile, nickel in Australia, and molybdenum (used for making hard steels) in the American West have been revealed by the new science of "metallogeny."

The Earth's Deep Interior

Other scientists (including coauthor Robert Hazen) devote their research lives to understanding more about the earth's deep interior. The mantle forms most of the solid earth, but we don't know its composition or temperature profile. We know the mantle convects, causing plates and continents to move, but not the details of how this process works.

Today's earth scientists approach these questions in two ways. One group, called mineral physicists, studies the properties of rocks and minerals that they subject to the very high pressures and temperatures that exist in the mantle. By learning in the laboratory how minerals respond to extreme conditions, they can identify which combination of minerals most closely matches the earth's deep interior. Mineral physics complements the work of the second group, seismologists, who increasingly focus on determining the three-dimensional structure of the earth. Seismologists once had to analyze signals from one earthquake at a time by hand. Today, however, supercomputers collate data on thousands of earthquakes, from hundreds of seismic stations around the world. Each piece of data places additional constraints on earth models. Eventually we hope to obtain a detailed three-dimensional picture of the convecting earth, a picture that will tell us as never before where our planet has been and where it is going.

Unstable Magnetic Poles

The earth's magnetic field owes its origin to the rotation of the liquid outer core, but beyond that we know little about why our planet behaves like a giant magnet. Since the core is electrically neutral, its rotation does not produce an electric current and cannot, in and of itself, produce a magnetic field. There are, however, somewhat more complex ways in which a rotating neutral conductor can create a field, so that's not the real quandary scientists face.

The real problem is that the earth's north and south poles have not always been where they are now. The magnetic pole wanders, usually remaining near the north pole, but changing position by a few miles every year. In addition, at various times in the past the earth's magnetic field has reversed, a process that may take a few hundred or thousand years while the "north" pole shifts to Antarctica. We can see over 300 of these reversals in the geological record, and we really don't have a very good idea of why they happen. Geologists are faced with the unenviable task of producing a theory that predicts a steady magnetic field that, at seemingly random times, flips directions.

The Formation of the Moon

The earth's solitary, lifeless moon provides a striking contrast to our dynamic planet. We may not know all of the details of the earth's formation, but we know enough to construct a reasonable scenario of its birth. As the material around the sun started to come together under the influence of gravity, the first thing that formed near the orbit of the earth were planetesimals—boulders anywhere from a few feet to a few miles across. As time went on, the planetesimals began to coalesce under their mutual gravitational attraction and the nascent earth began to grow. The bigger it got, the more material it attracted. Soon it was moving through its orbit, accumulating debris the way a car windshield accumulates bugs on a summer evening.

Viewed from the surface of the growing earth, this process must have been spectacular to behold. Huge meteorites would fall from the sky, blasting craters and heating the surface. These early craters have long since eroded away, but the surface of the moon (and of the moons of other planets) bears mute testimony to this time, which astronomers call the Great Bombardment. Eventually, the accumulated heat from these collisions raised the temperature of the earth to the point that rocks melted and heavy materials like iron were free to sink to the center. This process, called differentiation, was the origin of the core-mantle structure we now see in our planet.

The moon's origin poses a persistent problem in the theory of the earth's formation. The moon has roughly the same chemical makeup and density as the earth's mantle, and it used to be thought that it had been torn out of the earth after differentiation had taken place (in pre-plate-tectonic thought, the Pacific Basin was considered the moon's birth scar). Alternatively, some scientists argued that the moon formed elsewhere and was captured, fully formed, by the earth's gravity. For various reasons, neither of these theories works well. Today, the hottest theory is called the Big Splash. The idea is that after differentiation had taken place, one last giant moon-sized meteorite crashed into the earth, throwing a lot of mantle material into orbit. The moon then formed from this loose material by a process similar to the original formation of the earth.

CHAPTER 14

EARTH CYCLES

Next time you're at the beach, pick up a handful of sand and look at it—really examine it. You'll notice that each grain differs from its neighbors. Some may be black, others shiny, still others may be green or white or various shades of brown. If you look at the grains under a microscope, more differences appear. Some look smooth and rounded, some sharp and angular. All of these differences arise because the grains of sand you are holding, despite their differences, share one important property: all are part of one of the great cycles that operate on our planet.

The grains of sand are different colors because each comes from a different rock inland from the beach. The grains have different shapes because they may have been washed to the beach, buried, incorporated into new rocks, uplifted, and washed to new beaches many times. The great cycles of weathering and erosion of rocks, sedimentation, and creation of new rocks has gone on since the beginning of the earth, and will continue until the sun burns out and the planet dies. Because of this cycle, it is possible that you hold in your hand the very first grain of sand that formed on the very first beach when the earth was young.

As scientists examine nature in operation, they recognize many ongoing processes—natural actions that constantly change the surface of the globe. Rain falls, gradually washing away rocks and soils and creating sand and silt. Rivers flow, carrying those sediments from hills and mountains to the valleys below. Ocean and lake waters evaporate, creating new rain clouds. Rocks, water, and atmosphere—the matter that forms the outer layers of our planet—are forever being shifted from place to place.

Water evaporates from the oceans and flows back, sometimes on the surface, sometimes underground, and sometimes stopping for a while in an inland lake. When the climate turns cold, water is taken up in huge ice sheets that spread out from the poles, and sea levels fall around the world. With warmer weather the ice sheets retreat and the water flows back into the sea. Like rocks, water moves in cycles.

Even the air moves in stately cycles, from the prevailing winds that bring us our daily weather to the long-term effects that constantly change the climate.

In fact,

Everything on the earth operates in cycles.

Today, scientists recognize that all of the earth's cycles are connected, each influencing the others. We are beginning to see our planet as a kind of marvelous machine, full of turning gears and moving parts. And most wonderful of all, we are beginning to understand how that machine works and how all the parts fit together.

CYCLES OF CHANGE

No feature on earth is permanent. Mountains weather away, continents break apart, oceans disappear, glaciers form and melt. Change is the hallmark of our planet. Yet amidst all this change, there is constancy. For all practical purposes the earth has a fixed budget of atoms. For an atom to be used in one structure, it must be taken away from another. Like a child in a room filled with wonderful building blocks, the earth has a large but finite number of pieces to play with.

THE ROCK CYCLE

There are all sorts of rocks in the world. But despite this variety, geologists classify rocks into only three basic types: igneous, sedimentary, or metamorphic. Cataloging rocks is not just an academic exercise. Each type of rock records a different complex past—a past revealed by mineral textures and form. Each type of rock can be changed from one form to another and then back again. Geologists call these transformations the rock cycle.

Igneous Rocks

The crust of our planet began as molten rock; from space that early Earth must have appeared as a spectacular, incandescent ball. The rock cycle could not begin until that glowing outer layer began to solidify. In the beginning, all rocks on earth were igneous—fire-formed.

Volcanism is the most spectacular process that produces new igneous rocks today. Volcanic rocks arrive at the surface, either in air or under water, as magma—the molten form of rock. The most obvious and destructive volcanoes occur on land, where huge fountains of incandescent molten rock light the night sky and rivers of lava destroy life and property while reshaping the landscape. Most of those volcanoes produce dark basalt lavas, which possess a sticky fluid consistency before they harden. Occasionally, as in the Mount St. Helens eruption in 1980, lavas are thick and viscous like tar so that little flow can occur. An epic explosion may be required to relieve the pent-up pressure.

Many magmas fail to make it to the surface. Geologists call these molten masses that cool deep underground "intrusive" rocks, to distinguish them from more visible "extrusive" lava flows. Because they form deep in the earth, it may take many millions of years for the material above an intrusive rock formation to be lifted up and weathered away, so that the rock can finally appear at the surface. But given enough time, intrusive rocks can be uncovered to create prominent landmarks like

Mount Rushmore in the Black Hills of South Dakota, Stone Mountain in Georgia, and the highest peaks of the Colorado Rockies.

Devils Tower in northeastern Wyoming, which figured so prominently in the movie *Close Encounters of the Third Kind,* is a classic example of a volcano that didn't quite make it. The magma penetrated hundreds of feet upward through overlying sandstone, but it never breached the surface. Cooling in place, the intrusion developed long, vertical cracks as it contracted. Now, millions of years later, the soft sandstone has weathered away, leaving behind the spectacular plug of igneous rock with its graceful rock columns.

Sedimentary Rocks

Imagine the early earth. Jagged volcanic peaks rose from the steaming oceans, only to be battered and broken by wind and waves. Small chips of rock broke off and were washed down to the sea. Soon sandy beaches buffered land from sea. River valleys and lake bottoms gradually filled with sediments from the debris of weathered rocks. Given time, thick deposits of sediments—layer upon layer of igneous rock fragments—were themselves buried, baked, and turned to stone. Rocks formed in this way are called sedimentary rocks.

Weathering, the process that generates sediments by destroying rocks, takes many forms. Ocean tides, flowing rivers and streams, and windblown sand contribute to physical weathering, as does the wedge-like effect of water freezing in cracks and pores. Rocks can dissolve by chemical weathering, and they can be attacked by the actions of living organisms, the way that tree roots and grass can gradually break up a sidewalk. All of these processes provide the raw material for the formation of sedimentary rocks.

Over time, the accumulation of sediment may bury beach sand deep underground. Pressure and the heat of the earth's interior, together with minerals deposited by water, cement the grains together into a rock called sandstone. Later, mountain-building activity may lift this rock up and the weathering process will start again. Eventually each grain in the sandstone will be broken off and transported to another, as yet unimagined, beach. Many of the grains of sand at your favorite beach have been on other beaches in the distant past, and many will someday reappear on beaches of the future.

For the past three billion years, living organisms have themselves contributed directly to the formation of sedimentary rocks. Plants die and their stems, leaves, and trunks accumulate in swamps to form layers of coal, while microscopic organisms in the ocean die and contribute their skeletons to the accumulation of material at the bottom—material that will eventually become the sedimentary rock we call limestone.

Because they form from material filtering down to ocean and lake bottoms, sedimentary rocks usually appear to be layered. They often look like the pages of a book viewed end-on. Watch for them next time you're out driving. The most common sedimentary rocks are sandstone (made from sand), shale (made from silt and clay), and limestone (from the skeletons of microscopic organisms).

Metamorphic Rocks

Igneous and sedimentary rocks do not retain their original form forever. At the surface, they will be broken down by weathering to form new sediments. If they are buried, even more interesting changes can occur as the transforming effects of temperature, pressure, and time do their work. At high temperatures, clays and other common minerals give up water, like a brick baking in a kiln, while at high pressure, atoms in a rock rearrange themselves to form new and denser minerals, just as graphite converts to diamond if buried 100 miles down. Rocks that have been changed since they first formed are called metamorphic rocks.

These rocks tell incredible stories of the earth's unrest. Some New England rock outcroppings high on mountaintops contain minerals that could only have formed 20 miles beneath the surface, at temperatures near 1,000°C. Remnant rock structures reveal that those outcrops once lay at the bottom of a deep ocean basin, where their sediments were buried deeper and deeper as more continental material eroded off the land and accumulated in the sea. When an ancient collision of the North American and Eurasian plates crumpled and compressed those ocean basin sediments to form the Appalachian Mountains, deep-buried sediments subjected to the heat and stress of mountain building gradually changed. Layered limestone transformed to the marbles of Vermont, while shale turned first to slate and then to schist, a shiny metamorphic rock with big crystals of garnet and other high-pressure minerals. More than 200 million years of erosion and uplift have now brought those ancient rocks back to the surface to tell their tale, to weather and erode, and to begin the whole process anew. Humans contribute to the cycle by quarrying the marble for monuments, gravestones, and other transient reminders of earth's constant change.

Thus the cycle continues. All three forms of rock—igneous, sedimentary, and metamorphic—can weather to form more sediment, and all three can be subducted to melt or metamorphose and start the cycle anew.

THE WATER CYCLE

The earth holds only a finite amount of water, and virtually all the water at the earth's surface now has been there almost since the planet was born, yet it never seems to run out. The "ever-filled purse" that represents our water resource comes about because water, like rock, moves through cycles, constantly being used, constantly being replenished.

Any water that might have been on the earth's surface when it first cooled off would probably have blown off by meteorite impacts or the "solar wind"—the intense stream of particles emitted by the newborn sun. The water that now fills the oceans, as well as the gases that make up the atmosphere, must have waited out this early violent period in our planet's history safely stored in solid rock. Only later did they come to the surface, through volcanic activity.

Oceans and atmospheres are not an inevitable consequence of planet formation. Smaller worlds, like Mercury and the moon, are too small to retain any surface fluids. Fast-moving gas molecules like water vapor, nitrogen, or oxygen gradually escape the weak gravitational pull of these bodies. If the earth had been much smaller, there would have been no oceans and no life to enjoy them.

Our planetary reservoir holds almost 500 billion billion gallons of surface water in its oceans, lakes, rivers, ice caps, groundwater, and atmosphere. An unknown (but probably larger) amount is locked up in minerals in the crust and mantle, though this bound water is obviously not readily available for human use. The oceans account for more than 97 percent of the vast surface-water budget. An additional 2 percent is frozen in ice caps and glaciers, leaving less than one percent as usable fresh water. These percentages may change slightly, for example during ice ages, but fresh water will never account for more than a small fraction of the total supply.

The most familiar illustrations of the water cycle depict water evaporating from the oceans, forming clouds that rain on the land, and finally collecting in streams and rivers that return to the sea. This simple water cycle, which takes a few weeks or months to complete, is certainly a part of the story, but the complete cycle is much more complex. It involves many interlocking cyclical processes that occur over times from a few hours to millions of years. There are several key pieces to this global jigsaw puzzle.

The Oceans

Oceans cover three quarters of the earth's surface. Averaging about three miles in depth, the oceans are characterized by a thin surface layer (only a few hundred yards deep) that absorbs sunlight. That zone overlies a dark, cold reservoir in which almost all of the earth's water is stored. The deeper you go into the water, the colder and saltier it gets. The pressure also increases, reaching values of several tons per square inch in the deepest parts of the ocean. Near land, narrow bands of shallow water cover continental shelves—regions best thought of as flooded parts of the land rather than as parts of the ocean proper. They usually extend out a few tens of miles, where the bottom drops off abruptly into the abyss of the deep ocean.

The thin upper skin of the sea, called the "mixed layer," is special. Heated by the sun and well mixed with atmospheric gases, it teems with life, from microscopic plankton and algae to giant fish and sea mammals. In contrast, the deep ocean is dark and dense, subject to tremendous pressure and chilled to only a few degrees above freezing. This is why shallow waters like those of the continental shelves, the famous fishing grounds of Georges Bank in the North Atlantic, and Chesapeake Bay produce so much seafood. The deep ocean, on the other hand, is a kind of desert, inhospitable and spare of life.

The deepest ocean waters circulate slowly and can spend many thousands of years in the dark void. Much of the densest, deepest ocean water comes from melting Antarctic ice; that ice water descends to the bottom, spreads out, and slowly rolls across the ocean floor to places as far away as the Bering Strait of the North Pacific.

Virtually all interactions between the oceans and other parts of the water cycle take place at the ocean surface. Rivers and rain add to the top layer, while surface evaporation returns water to the air. Surface waters also provide coastal areas with a thermal buffer, moderating air temperatures during the coldest and hottest months.

Ice Caps and Glaciers

At times in its history the earth has had as much as 5 percent of its water budget tied up in ice caps and glaciers, at other times as little as a fraction of a percent. The amount of earth ice depends in a complicated way on the positions of the continents, and in a regular and predictable way on variations in the orbit of the earth around the sun and the direction of the earth's axis of rotation. At present, during a period of moderate temperatures, about 2 percent of the globe's water (about three quarters of all fresh water) is frozen.

The largest concentration of ice on earth today is in the thick glaciers that cover the south pole. Such large ice caps can accumulate only if a continent covers one of the poles, giving a base of solid ground to support thick ice. Otherwise, we have a situation like the one at the north pole today. As the northern ice cap starts to build up, the weight of new ice pushes old ice deeper into the water, where it melts because of the higher pressure. Thus, without a solid support, an ocean ice cap can never be more than a few hundred yards thick, and can never contain more than a tiny fraction of the earth's water. For three quarters of the earth's history, there were no continents at the poles, and hence no large ice caps. Occasionally there have been continents at both poles, a condition which probably led to more water being locked up in glaciers than at present.

Scientists have trouble predicting continental wandering, not to mention the effects of those movements on glaciation, but purely astronomical effects on ice caps are easy to predict. The most important effect involves the tilt of the earth's axis of rotation, which now leans 23 degrees off the axis of the earth's orbit. Northern and southern hemispheres have opposite seasons at present because when the northern hemisphere is tilted toward the sun (in summer), the southern hemisphere is tilted away. In general, any effect that makes summers cooler contributes to glaciation. The reason is simple: if summers are cooler, ice and snow will stay on the ground longer in Canada and Siberia. This ice and snow will reflect sunlight, further lowering the temperature and allowing still more ice and snow to remain on the ground the next year. The result: over a period of a few thousand years, large sheets of ice spread out from the pole and the high mountains and cover large parts of Europe and North America. When this happens, we say there is an ice age.

The most recent ice age was in full swing 20,000 years ago,

when glaciers extended as far south as Chicago. Today we live in what geologists call an "interglacial," a term we find chilling in every sense of the word. On a shorter time scale, chance events also affect the glacial cycle. Major volcanoes can spew dark matter into the atmosphere, blocking sunlight and cooling the planet for a year or two, which creates brief periods of growing glaciers. Increased atmospheric concentrations of carbon dioxide and other "greenhouse gases," whether man-made or natural, may have the opposite effect. The best prediction available now is that we are (or should be) heading into another period of glaciation, although the advent of global warming by the greenhouse effect may introduce a temporary glitch in the grand cycle of freezing and melting.

The earth's glacial cycles are of more than abstract interest. If more water is tied up in glaciers, less remains to fill the oceans and sea levels will fall. During the last glaciation, for example, the eastern coast of North America was 150 miles farther east than it is today, and on the west coast an ice and land bridge existed between Siberia and Alaska. Anthropologists believe this bridge allowed the ancestors of the American Indians to reach this continent. Conversely, geological evidence also points to warm interglacial periods, some within the past hundred thousand years, when oceans were 100 feet higher than today. To see what such a sea level change would mean, imagine the present New York and Los Angeles waterfronts under 100 feet of seawater.

Fresh Water

The underlying principle of the science of hydrology, which concerns itself with the study of the water cycle, is that water is a mobile resource. It flows by gravity from high to low, thus shaping the land and playing a central role in the rock cycle. It evaporates from land and sea into air, forming clouds and playing a major part in the weather cycle. It falls as rain on the land, filling reservoirs with fresh water and providing the chemical medium for life.

Only a small fraction of the earth's liquid resources are available to land plants and animals as fresh water. Rainwater that falls to land can follow many different paths. Some water penetrates the soil, some enters lakes, ponds, and streams, some evaporates quickly and returns to the clouds, but most (as much as 99 percent of the total) becomes part of the vast underground reservoir of groundwater. Groundwater accumulates in aquifers—porous rocks such as sandstone—where continuous networks of tiny spaces between mineral grains form huge reservoirs. Water soaks in wherever the aquifer is exposed at the surface, and the reservoir fills by the force of gravity. Aquifers, bounded top and bottom by impervious rock layers, can be tapped by deep water wells. It can take many thousands of years for an aquifer to fill up with water, so tapping that water is analogous to mining a mineral deposit. Throughout the American West, wells are going dry as the supply of stored water is drained. The fact that they will be replenished in a few thousand years is of scant comfort to ranchers and farmers.

Humans, who depend on the natural water cycle for their survival, affect the cycle in many ways. We divert streams for crop irrigation, the largest single use of water. We dam rivers for hydropower. We create artificial lakes for water storage and recreation. We use flowing water to purge unwanted chemicals from our factories and sewage from our homes.

Until fairly recently, humans regarded fresh water as an inexhaustible resource, one that could be used with little regard for long-term consequences. To some extent this attitude was justifiable, since some aspects of the cycle are resilient. Evaporation can purify water in a relatively short time. We can clean up polluted rivers, streams, ponds, and coastlines in a few years, as newly evaporated water replaces the old, polluted fluid. But there are other problems less amenable to short-term solutions. Contaminated groundwater may remain polluted for decades, thus diminishing our stock of potable fresh water at the same time that growing populations demand more. Politicians, as well as scientists, now realize that humans are an important part of the earth's water cycle.

THE ATMOSPHERIC CYCLE

People instinctively distinguish three principal atmospheric cycles. Weather is the short-term cycle, somewhat unpredictable and almost always different from the "average" weather for a particular day or month. The longer-term cycle related to the movement of the earth around the sun we call the seasons, and we use them to number human lives and accomplishments. On a much longer time scale we recognize the cycle of climate. The climate of a region rarely changes significantly in a single lifetime, but the passage of several generations may be sufficient to alter the severity of winters, to turn productive farm land into desert, or to transform swamps into firm ground.

Weather, seasons, and climate all involve the earth's atmosphere, an envelope of gas surrounding our planet that is, in its way, as complex as the ocean and the crustal rocks. To understand the weather you must understand how the atmosphere is constructed. The atmosphere behaves in many ways like the solid earth and its oceans. Like hot rocks in the mantle or currents in the sea, air circulates. And like the solid earth and the oceans, the atmospheric system has layers that differ in temperature and pressure.

The troposphere is the warm layer of air next to the surface of the earth. It extends up about 40,000 feet, and provides the expressway for commercial jet travel. The troposphere, high enough to cover Mount Everest, contains most of the cloud systems we see from earth, but large thunderstorms often produce clouds that stick out above it. Above the troposphere lie successive layers called the stratosphere (to 150,000 feet), mesosphere (to 260,000 feet), and the ionosphere, each of which plays a role in the overall behavior of the ocean of air in which we live.

Convection and the Weather

Convection, the same mechanism that drives plate tectonics, causes weather in the near earth layers. The lower atmosphere is a giant convecting system, powered by solar energy and wrapped around the turning planet. Most of the sun's energy arrives near the equator, where warm air expands and rises. On a nonrotating planet, rising air would circulate from equator to pole in high atmospheric currents, eventually to cool and descend near the poles. The surface flow of air would travel from pole to equator. In the northern hemisphere of such a world, wind and weather would usually come from the north.

The earth's twenty-four-hour rotation complicates this picture by stretching the simple north-south convection cell into three large east-west convection cells. Surface winds blow west to east in the temperate regions of the northern and southern hemispheres, creating the "prevailing westerlies." One of these cycles dominates the west-to-east weather pattern characteristic of most of the United States. In equatorial regions the pattern is reversed with east-to-west "easterlies," such as the Atlantic Ocean's trade winds that provided sailors a speedy voyage from Europe to the Americas. Stagnant air masses, such as the "doldrums" at the equator, form in between these major air currents.

Weather Report Jargon

No earth system is more carefully monitored than the lower atmosphere. Radio and TV news programs provide frequent updates on present values and probable changes in several key atmospheric properties: in particular temperature, pressure, humidity, wind speed and direction, and pollution. Other terms, such as "wind chill factor" and "comfort index" combine two or more of these basic variables to indicate how good or bad it feels to be outside.

Temperature variations in space and time are often extremely complex. Not only does the temperature rise and fall on a daily and seasonal basis, but temperature also varies by dozens of degrees with altitude. Forecasters report surface temperatures, yet much of the weather action—rainfall, for example—depends on the very different temperatures of upper level air.

Barometric pressure is the weight of all the air overhead—roughly fourteen pounds per square inch (which is the same weight exerted by a column of mercury about 30 inches high at sea level). On a planet with perfectly still air the pressure would hardly vary at all. But on our world large air masses form vast circular currents, which pile up air at their margins (high-pressure zones, or "highs") and suck air out of their centers (low-pressure zones, or "lows").

Humidity is a relative term, reported as a percentage. The amount of water vapor that air can retain without producing fog or rain varies greatly, and depends on air temperature. On a 90-degree day the atmosphere can hold several percent water by weight, while in midwinter New England air holds less than a half percent water. The relative humidity is a measure of how much water the air actually holds, compared to how much water the air can absorb. When the relative humidity is high, perspiration cannot evaporate easily from your skin and you feel uncomfortable. As the temperature drops, the air's ability to hold water decreases, and water condenses out in droplets. That's why water condenses on the outside of your iced drink glass even on relatively dry summer days.

Wind speed and direction in North America depend on many factors that modify the general west-to-east flow. Topography, the location and temperature of large bodies of water, and the distribution of pressure highs and lows all play important roles. In recent years it has become fashionable in some television weather circles to provide fancy computer graphics showing the meandering path of the jet stream. Jet streams are high-altitude wind currents that behave like fast-flowing rivers of air about 8 miles high. Roughly speaking, the jet stream divides cold northern air from the warmer air masses in temperate regions. The general jet stream trend is always west to east, but like rivers they adopt sinuous paths, changing position and speed on a daily basis, influencing and being influenced by the location of surface highs and lows. Because of their great speed, often greater than 100 miles per hour, jet streams may affect aircraft operation and scheduling. Flights from New York to California typically take an hour longer than the return because of this effect.

Natural and man-made pollutants have become a fixture of weather reporting. Concentrations of airborne pollens, especially prevalent during spring and summer months, are described by an arbitrary scale of particles per given volume of air. Most city weather reports include similar statistics on smog, a collective term for a variety of man-made pollutants.

FRONTIERS

Scientists traditionally studied the earth by looking at each of the great cycles in isolation. Today the focus is changing, and researchers are becoming more and more interested in seeing the entire earth as a single system, with each cycle affecting, and being affected by, the others. There are plans, for example, to launch a fleet of satellites to study the ocean, the atmosphere, and the landforms of our planet and, most important, to coordinate the data obtained. If this project is carried through to its conclusion, by the year 2000 we will have the beginnings of a data base that will serve as a benchmark against which to measure future changes.

At the moment, the main effort in "intercycle" research is concentrated on the interaction of the atmosphere and the ocean. Oceanographic ships routinely measure gases passing through the ocean surface. Such research typically seeks to find out how much carbon dioxide is moving into the sea—a number that has considerable impact on our predictions of the consequences of global warming by the greenhouse effect.

As our computer models of the atmosphere become more sophisticated, it becomes important to test them against real data, especially if these models are going to serve as guidelines for political action. Unfortunately, we only have detailed data for the present-day earth. We know that in the past the continents were in different locations and the weather was quite different from today's. If our climate models could describe these earlier earths correctly, we would have a great deal more confidence in our predictions than we do now. Thus, we find ourselves in the paradoxical situation of needing information about the climate millions of years ago to predict what the climate will be in the next century.

CHAPTER 15

THE LADDER OF LIFE

Imagine lying out on a hill on a warm summer afternoon. All around you is evidence of the physical earth—rocks, clouds in the sky, perhaps a distant river or lake. But you are also surrounded by evidence of a different earth. The grass under your back, the insects you hear buzzing, the birds wheeling in the sky, and you yourself are all parts of a great web of living things that surround our planet. The web extends underneath the soil, to the depths of the ocean, to deserts and forests. It feeds you, supplies the air you breathe, and makes your life possible. And despite its diversity— despite the difference between a blade of grass and a giraffe— everything in this great web is related to everything else.

The most striking verification of this relationship is in your chemistry. Your body contains the same chemical compounds, derives its energy from the same chemical reactions, and utilizes the same chemical mechanisms as every other living thing on the planet. At the very core of your being, you are part of the earth's life.

You can think of life as arranged in a great ladder, starting with the basic chemical compounds that make up living things, pro- gressing upward to cells, then to collections of cells that make up organs, organ systems, and finally organisms themselves. The cells, midway up this ladder, are the nexus of life, a fact that can be summarized as:

All living things are made from cells, the chemical factories of life.

Cells act as chemical factories, taking in materials from the environment, processing them, and producing "finished goods" to be used for the cell's own maintenance and for that of the larger organism of which they may be part. In a complex cell, materials are taken in through specialized receptors ("loading docks"), processed by chemical reactions governed by a central information system ("the front office"), carried around to various locations ("assembly lines") as the work progresses, and finally sent back via those same receptors into the larger organism. Far from being a shapeless blob of protoplasm, the cell is a highly organized, busy place, whose many different parts must work together to keep the whole functioning.

THE MOLECULES OF LIFE

Modules and Shape

Each of the many different kinds of molecules that make up living things on our planet is built from an exact and orderly arrangement of atoms. Two important features characterize all of these molecules: (1) all are made from a few small modules, and (2) their properties depend mainly on their shapes.

No matter how big and complex an organic molecule gets (and some can contain millions of atoms), its basic structure is always relatively simple, strung together from a few basic components. Every building, from a country cottage to the Sears Tower, is a different arrangement of common elements like bricks and windows. Similarly, you can think of every organic molecule, from simple sugars to complex proteins, as an assembly of simple building blocks. There are four main types of organic molecules. Each is built in a different way from different components, but they all share this basic property of modularity.

When molecules become large and complex, their three-dimensional shape becomes important in a way that doesn't apply to relatively simple compounds like the minerals we described in Chapter 7. No matter how complex a molecule is, it must interact with other molecules by the way of the same types of chemical bonds that hold smaller molecules together. Ultimately these bonds depend on the interactions between electrons of two neighboring atoms. Thus, in order to produce reactions between complex molecules, the appropriate atoms in each of the participants must be brought near each other so that their electrons can interact. Two complex molecules can interact only if their shapes exactly match, like two pieces of a jigsaw puzzle.

You can think of two large molecules that can interact with each other as long ropes haphazardly thrown into piles. Think of the atoms that want to form bonds as patches of Velcro, several on each rope. The ropes will not stick together if you press any two pieces at random; the probability of forming a bond by tossing one pile on top of the other is pretty small. A strong bond forms only if you arrange the ropes in such a way that the patches of Velcro can meet each other.

There are many ways that the piles of rope might not match. It might be, for example, that a patch on one rope is hidden inside a deep recess. In this case, only a rope with its Velcro at the end of a long loop could make proper contact. In the same way, complex molecules can be made to bond only if they fit properly. Therefore the interactions of these molecules depend on their shapes.

Enzymes

Enzymes are an extremely important class of large molecules whose sole task is to help other molecules link together. In this linking process the enzymes themselves remain unchanged. (Molecules that facilitate such reactions between *inorganic* molecules are called catalysts.) In general, enzymes are remarkably specific—they link only two particular kinds of molecules and no others.

If you took the two piles of rope in our example and twisted them around so that all the Velcro pieces stuck together, you would be acting as an enzyme. You would grasp one rope, then the other, in such a way that the sticky patches faced each other. Once the ropes were stuck together, you would walk away unchanged, ready to work your magic on the next pile of rope.

Thus does an enzyme bring two specific molecules near each other, let their atoms form bonds, and then, its job done, go on its way unaffected, ready to begin the process all over again. Each kind of enzyme supervises one kind of reaction. For each of the thousands of chemical reactions that go on in each cell in your body every day, there has to be a separate molecule to act as an enzyme.

The Role of Carbon

Large molecules must be constructed from atoms that are both plentiful and easy to hook together. Carbon, with four electrons in its outermost orbit, has both of these qualities. It also possesses an even more important property, the ability to form strong covalent bonds with other carbon atoms. Thus, it is possible to put together long chains of carbon atoms with each atom in the chain having free electrons that can act as "hooks" to form covalent bonds with other atoms. This property of carbon explains why it is found in all molecules in living systems, and why we say that life on earth is carbon-based. We might, incidentally, also call it covalent-based, since by far the greatest number of bonds between atoms in organic molecules are of this type.

In addition to carbon, organic molecules often contain five other types of atom—hydrogen, nitrogen, oxygen, phosphorus, and sulfur. (A simple mnemonic—CHNOPS—helps to keep these atoms in mind). From these six elements we can construct all the basic modules needed to assemble the organic molecules themselves. Thus, on the scale of the atom, a basic simplicity underlies the diversity of life on our planet.

The Four Molecules of Life

Four types of molecules are essential to the working of a cell. They are:

1. *Nucleic acids*: These molecules (DNA and RNA) carry the blueprint that runs the cell's chemical factories, and also are the vehicle for inheritance—the passing of genetic information from one generation to the next. Their structure and function are discussed in the next chapter.

2. *Proteins*: Proteins are the workhorses of the cell. In addition to their familiar role in the structure of living things (your hair and fingernails are made from protein, for example), proteins serve as almost all of the enzymes that run chemical reactions in cells. Life would not be possible without these molecules.

Proteins are modular, built from hundreds or thousands of smaller molecules called amino acids. The basic structure of an amino acid consists of a group of hydrogen and nitrogen atoms, on one side, a group of carbon, oxygen, and hydrogen on the other, and a "side group" labeled R in the diagram. There are hundreds of possible choices for this side group, each corresponding to a different amino acid.

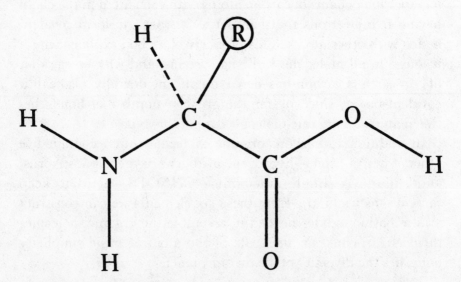

Amino acids link together in long chains to form proteins. Twenty different amino acids, distinguished by different groups of atoms in the position "R," are found in common proteins.

Amino acids link together by having the H on one end of one molecule combine with the OH at the end of another molecule to form water (H_2O). Think of this as the two molecules squeezing out a drop of water as they cement themselves together. Once two amino acids are linked in this way, a third can be added on, then a fourth, a fifth, and so on. Each different chain-like sequence of amino acids results in a different kind of protein molecule. Proteins can range in size from a few hundred amino acids (insulin is an example of this type of small protein) to giant chains containing hundreds of thousands of links.

Once an amino acid chain forms, it can take many shapes. The chains can coil up into a corkscrew (as they do in your hair), or they can wrap around each other to form a cable (as they do in the tendons that hold your muscles in place). Very long proteins may even have different structures along different parts of the chain. Once this so-called secondary structure has been established along a chain, large proteins can also fold up into complex, irregularly shaped globules. The nooks and crannies in the surfaces of these complex folded proteins make them ideal for use as enzymes in the cell. The smaller molecules that serve as the raw material for the cell's chemical reactions fit into the surfaces of specific proteins.

One strange and as yet unexplained fact about the structure of proteins in living systems has important implications for evolution. There are hundreds of possible amino acids, but only twenty different kinds are actually found in the proteins of living things on earth.

3. *Carbohydrates*: While proteins supervise the cell's chemical factories, carbohydrates provide each factory's fuel supply. The basic building blocks of carbohydrates are sugars—small ring-like molecules with perhaps two dozen atoms of carbon, oxygen, and hydrogen. Common sugars include glucose (an ingredient of many hospital IV solutions), fructose (found in many fruits), and sucrose (ordinary table sugar) made from a ring of glucose and a ring of fructose bound together.

Like the amino acids in proteins, sugars can "squeeze out water" by combining an H from one ring and OH from another, thereby forming a bond that holds the rings together. And, as was the case for proteins, these simple building blocks can be hooked together ad infinitum to form long chains. Strings of glucose molecules stacked head-to-tail in slightly different ways produce both starch and cellulose, two large molecules that are very important in the architecture of living systems. Starches store energy in cells, while cellulose is the principle fiber that stiffens the structure of plants.

The fact that the cotton shirt you're wearing, the starch in your muscles that allows you to put a little extra effort into today's run, and the sugar you put into your morning coffee are all made from the same basic building blocks is a good illustration of the diversity that can result from the simple modular structure of organic molecules.

4. *Lipids*: "Lipid" is a catch-all classification of molecules, including any organic molecule that doesn't tend to dissolve in water. Think of lipids as fats and oils, like the little glossy drops that float on top of your soup. One important class of lipid molecules consists of a long chain in which one end is attracted to water, the other repelled. Lipids play a variety of roles in living systems. They are very efficient at storing energy: that little bit of extra weight around your middle that you really are going to lose one of these days is made of lipids, for example. More important, because lipids do not dissolve in water they are ideal materials for cell membranes, both those that separate the cell from its environment and those that delineate separate structures inside the cell.

One property of lipids that you're likely to encounter has to do with how the carbon and hydrogen atoms in the molecule bond together. When neighboring atoms share two electrons, the bond is said to be "saturated"; if only one, the bond is "unsaturated." Electrons in saturated bonds are easily broken off, so saturated molecules combine relatively easily with others. In general, animal lipids are saturated, plant lipids (like olive oil) unsaturated.

Many health problems in America arise from overconsumption of saturated fats.

THE CHEMICAL FACTORIES OF LIFE

Cells are the basic unit of life. Many living things, in fact, consist of only one cell. Others are multicelled—your body, for example, contains about 10 trillion cells. Cells come in a wide variety of shapes and sizes. The largest cell (an ostrich egg) is bigger than most animals, and the smallest (certain bacteria) can barely be seen with the strongest light microscope. Most cells are about a ten thousandth of an inch across, a bit smaller than the particles of smoke that make the sky hazy after a fire.

Cells have two primary functions: to provide a framework that supports the complex chemical reactions required to sustain life, and to produce exact copies of themselves so that the organism of which they are a part can go on living even after those cells die. In this chapter we focus on the cell as a chemical factory; in the next, on how cells reproduce.

Like any factory, each cell has several essential systems. It must have a front office, a place to store information and issue instructions to the factory floor to guide the work in progress. It must have bricks and mortar—a building with walls and partitions where the actual work goes on. Its production system must include the various machines that produce finished goods as well as the transportation network that moves raw materials and finished products from place to place. And, finally, there must be an energy plant to power the machinery.

The Front Office

In each cell of your body a nucleus acts as the front office. Separated from the rest of the cell by a double membrane, the nucleus keeps the nucleic acid DNA on file. DNA is something like an instruction manual. The DNA manual itself can't do the work, but it contains information that programs the cell to carry out its functions.

Primitive cells do not have nuclei, but carry their DNA instructions loose inside their cell wall. Such cells are called "prokaryotes" ("before the nucleus"). More advanced cells, including all those in multicelled organisms like human beings, isolate the DNA in a nucleus. Such cells are called "eukaryotes" ("true nucleus").

Bricks and Mortar

Cellular factories consist of walls, partitions, and loading docks. Cell membranes provide all interior and exterior walls. A typical cell membrane is made of lipid molecules arranged in a double layer, with water-repellling ends head-to-head on the inside and water-attracting ends on the outside.

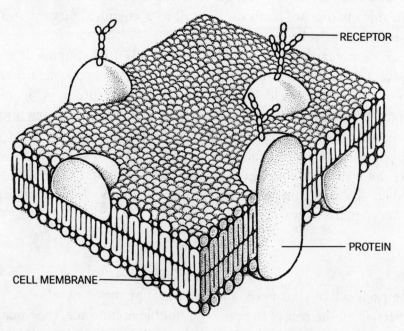

A typical cell membrane consists of a double layer of lipid molecules, interspersed with large protein molecules that act as receptors.

Here and there on the membrane surface are large protein and carbohydrate molecules, which have complex three-dimensional shapes and act like specially designed loading docks. These "receptors" lock onto only one specific kind of molecule in the outside environment. When a receptor "recognizes" its specific molecule (perhaps a sugar or an amino acid), it binds to it. A common outcome of this binding is for the external object to be sucked into the cell's interior, where it is surrounded by a small bit of membrane called a vesicle. The vesicle then becomes the vehicle in which the material is transported around the cell. The reverse process, in which a vesicle approaches the cell wall from the inside, joins to it, and dumps its contents outside the wall, is the primary mechanism by which a cell returns materials to the environment.

Your cell's loading dock system can sometimes be fooled, with tragic consequences. The AIDS virus, for example, happens to fit the receptors normally found in the membrane of the human white blood cell. Thus, once taken into the body, the virus readily enters those cells and kills them, destroying the person's immune system in the process.

The enfolding of external objects by a cell membrane is thought to explain the existence of the nucleus as well as some other structures inside the cell that you'll meet shortly. Scientists think that early in life's history all cells were prokaryotes—none had nuclei. At some point one cell swallowed another, forming a new symbiotic system. Double membranes that now surround the nucleus and many of the other parts of your cells provide a telltale indicator of this process. Any engulfed cell carries its own membrane as well as the part of the original cell membrane involved in the engulfing process. Thus, many biologists think of eukaryotic cells as evolved colonies of simpler cells, with each part contributing to the whole.

The Production System

Production areas that carry out the chemical work are themselves separated from the rest of the cell by membranes. These structures, and any other organized body within the cell, are called organelles.

Each organelle performs a specific chemical function. Some supply the cell with energy, some digest the molecules that serve as food, some produce proteins necessary for the operation of the cell, some put finishing touches on molecules that have been created elsewhere in the cell, and some serve as stable platforms on which the process of protein assembly takes place. At any given moment, all of these functions operate at different places within the cell.

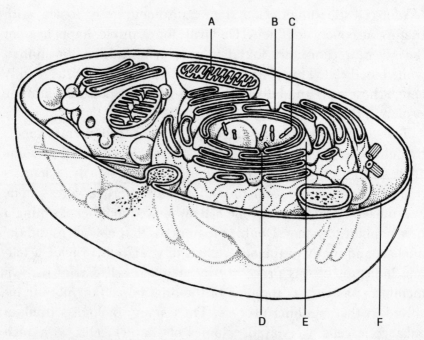

A MITOCHONDRION   D NUCLEAR PORE
B NUCLEAR MEMBRANE E NUCLEUS
C CHROMOSOME   F CELL MEMBRANE

Every living thing is composed of one or more cells, each of which has a complex anatomy. A "generic" cell contains many structures and organelles — tiny chemical factories.

Each organelle has its own complement of receptors. Materials are moved across the boundaries of organelles, and from one place to another within the cell, in vesicles of the type described above. Inside the cell, connecting all its parts, a complex web of fine filaments serves as the streets and highways of the cell's transportation system. At any instant, thousands of cargo-laden vesicles, carrying raw materials and finished products, are whizzing around on these filaments from one part of the cell to another. Organic molecules in the membrane of the vesicles fit into receptor molecules in cell membranes, guaranteeing that the load gets delivered to the right place every time.

The Power Plant

Cells, like any factory, need energy to operate. Two different kinds of cellular power plants have evolved. Some cells absorb energy directly from the sun, while others gather energy by eating other organisms that have stored it. Plants use the first strategy, animals the second, and the structure of their cells reflects this difference.

Plants acquire energy directly from sunlight through the process of photosynthesis. In this process, molecules of chlorophyll or related pigments absorb photons from the sun. The photons' energy is converted into chemical energy that the plant can use to grow and reproduce. In the course of this rather complicated chemical process, carbon dioxide and water from the cell's surroundings are converted into glucose (or other carbohydrates) plus oxygen. The net effect of photosynthesis, then, is to remove carbon dioxide from the air, produce energy for the cell, and give off oxygen as a waste product.

Animals, unlike plants, cannot convert the sun's energy directly to food, and therefore must get theirs by eating plants or by eating animals that eat plants. The food you eat contains energy in the form of the bonds that hold its molecules together. After the food has been broken down, it is taken into the cells where its energy is released by a process called respiration. Think of respiration as a slow burning. It allows molecules like glucose to combine with oxygen, releasing the energy tied up in the molecular bonds in the process. Its waste product is carbon dioxide, which you breathe out.

Photosynthesis and respiration are complementary. The carbon dioxide you breathe out is used by plants to create glucose; the oxygen that plants give off in turn serves as the raw material for respiration. This eternal cycle between plants and animals is a key feature of ecosystems on our planet.

The simplest single-celled organisms use fermentation, a more primitive and less efficient method of burning their fuel that does not require the presence of oxygen. The so-called anaerobic bacteria responsible for turning a pile of garbage into humus operate in this way, as does the yeast that ferments grape juice into wine. Presumably all life generated energy by fermentation in the early Earth, when there was no oxygen in the atmosphere.

More advanced cells with nuclei often retain the ability to produce energy by fermentation as a backup to the more efficient process of respiration. Strenuously working your muscles, for example, may deprive them of oxygen, causing your cells to fall back on fermentation, which in humans produces lactic acid as a by-product. The buildup of this acid makes your muscles sore the next day.

The cell's energy is generated in tiny sausage-shaped bodies called mitochondria. Partially digested carbohydrates, fats, and proteins from food are ferried to the mitochondria, where they are "burned" to produce energy to run the cell. A typical cell has hundreds of mitochondria at work.

Energy extracted from glucose or other fuel in a mitochondrion is usually not used immediately, so the cell must have a way to transport the energy from the place where it is produced to the place where it is to be used. To carry out this job, the cell uses a variety of molecules that serve as energy carriers. They play the same role in the cell that cash plays in an economy, allowing energy created in one place and time to be "cashed in" at another place and time, thereby making the working of the cell possible.

The "energy coin" molecules all work in the same way. The cell hooks electrons or groups of atoms to the molecule in one place, which requires energy. The molecules then move to another place where the electron or group of atoms is unhooked, releasing the stored energy. Almost all cells use a molecule called ATP (adenosine triphosphate) for relatively small amounts of energy, and a variety of other molecules to transport larger amounts when necessary.

Thus the cell is a dynamic, bustling place involving thousands of individual parts, each carrying out its own specialized chemical function to support the effort of the whole.

THE ORGANIZATION OF LIFE

When you studied biology in school, chances are you spent a lot of time learning the names of organisms and their parts. This book doesn't do that, because detailed nomenclature, while essential for describing individual beasts and flowers, reveals little about general biological principles and is unlikely to be the subject of public debate or news reports. Taxonomy—giving things names—is important to science but is not essential to achieving scientific literacy.

Living things may consist of one cell only, or they may be multicellular. In a one-celled organism, the process of energy acquisition and the production of chemicals must all go on within each cell. In more complex organisms, like humans, cells specialize and the work is divided. The cells are organized into organs, and the organs into organ systems. Your digestive system, for example, contains a trillion cells, each its own complex chemical factory, each designed to do its bit in the system that takes in food and converts it into raw materials for use in the rest of your body. A combination of organ systems makes a complete organism.

Finding a way to order and catalog the wonderful diversity of living things was, until this century, one of the main tasks of biology. The general scheme used today grows out of the classification first proposed by the great Swedish botanist Carl Linnaeus (1707–78). Think of every living thing as being situated somewhere on the branches of a great tree. The Linnaean classification locates a specific organism by specifying the main branch, then the side branch, then a smaller side branch, and so on until we arrive at the final twig on which the organism is found.

Aside from occasional regroupings (particularly of fossils), the job of classifying living things is no longer considered to be a major research area in biology. Apart from medical research, neither is the study of complete organisms and their internal organs a major focus of modern biological sciences. For the most part, the study of cells, their constituents, and the molecules that compose them is where the action is today.

The Five Kingdoms

Until very recently, schoolchildren learned to classify every living thing as either a plant or an animal. Today biologists recognize that the tree of life is better described as having five great trunks, each called a kingdom. In addition to animals and plants, the kingdoms include fungi, single-celled organisms with a nucleus (or protista), and single-celled organisms without nuclei (or monera).

Each kingdom is further divided into increasingly specialized branches called phylum, class, order, family, genus, and species. The human being, for example, is a member of the animal kingdom, phylum of chordates (animals with spinal cords), subphylum of vertebrates (chordates with backbones), class of mammals (vertebrates with hair who suckle their young), order of primates (mammals with opposable thumbs and big brains), and family of hominids (primates that walk erect and have other distinctive skeletal features). All members of this family except for genus *Homo*, species *sapiens,* are extinct. Human beings, therefore, are referred to as *Homo sapiens* in the Linnaean scheme. The farther we move back toward the main trunk in this scheme, the more inclusive the categories become—rabbits are mammals but not primates, fish are vertebrates but not mammals, and so on.

The end product of this classification scheme is the species, which biologists define to be a single interbreeding population. All human beings can interbreed, for example, so we are all members of the same species. Polar bears and black bears, on the other hand, both members of the genus *Ursus,* do not interbreed, and so are different species. We commonly describe organisms by two Latin words—like the dinosaur *Tyrannosaurus rex* or man's best friend *Canis familiaris*—that represent the genus and species, and the entire classification scheme is implied by those names.

Single-celled Organisms

We now give single-celled life forms two entire kingdoms on the tree of life, depending upon whether they have a nucleus. Here's how several familiar one-celled creatures fit into the modern scheme.

Bacteria, a phylum of monera, are usually named according to their shape: cocci (spheres) or bacilli (rods). They cause a number of diseases, like syphillis, tuberculosis, and cholera, but also are instrumental in producing many antibiotics. Anaerobic bacteria get their energy from fermentation and can survive without oxygen. They decompose much of the world's organic garbage.

Plankton include any small organism that floats on the surface of water. The most abundant plankton are bacteria, referred to loosely as "blue-green algae," which produce much of the world's oxygen supply.

The familiar amoeba is not just a free-form blob, but a single-celled organism with a nucleus and a complex internal structure. It moves around, engulfs its food, and is a favorite creature in high school biology labs.

FRONTIERS

Protein Structures

The key to understanding how protein molecules (enzymes) perform their chemical tasks lies in their complex three-dimensional shapes. The bumps and valleys on an enzyme's surface serve to attract and then connect other chemical components. Each enzyme is thus a precise arrangement of thousands of atoms—primarily carbon, oxygen, nitrogen, and hydrogen. Protein crystallographers devote their research lives to unraveling the locations of all those atoms, in the hope of understanding the chemical basis of life. Deducing a protein structure is no easy task. It can take years of research to unravel the structure of a single modest-sized protein molecule.

We now know the structures of such important molecules as hemoglobin, chlorophyll, and insulin, but there are thousands of other complex proteins in living things. Protein crystallography will remain an important research discipline for many decades to come.

Neurobiology

The nervous system of an animal is not a simple electrical circuit. When a signal gets to one end of a nerve cell, the cell sprays various molecules out for the next cell to pick up. Thus the transmission of a nerve impulse is a complex chemical, as well as electrical, phenomenon. There is an enormous research effort under way aimed at understanding nervous systems and how single nerves and their connections lead to larger-scale phenomena like behavior and learning. Ultimately, the goal is to understand the working of the human brain, but in the meantime scientists are content to work on single nerve cells (like the axon of the giant squid) and on small, relatively uncomplicated creatures like worms and cockroaches.

Immunology

Like all other vertebrates, you possess an immune system designed to defend you against foreign cells and molecules. The main components of your immune system are five different types of white blood cells, each of which has a specialized role to play. One type (the so-called B cells) produces antibodies, which are Y-shaped molecules in which two ends of the Y fit specific kinds of foreign molecules. Antibodies thus lock onto the unwanted object. The third leg of the Y-shaped antibody then stimulates other parts of the immune system to destroy the entire antibody-plus-foreign-molecule package. This part of the immune system protects against small invaders like toxins and some bacteria.

Other white blood cells (called T cells) contain receptors that recognize molecules on the surface of foreign cells. The T cells then bind to the foreign cells and destroy them. In this way, your body eliminates parasites and cells altered by cancer or viruses. Other kinds of T cells regulate the actions of the immune system. Some cause antibodies to be produced after the first exposure to a new invader; this is how we acquire immunity to diseases like measles. Still others suppress the action of the immune system.

The immune system is being studied intensely today, primarily because of its importance in medical research and treatment. The AIDS virus, for example, destroys T cells that regulate the immune system, which then loses its ability to respond to new diseases or eliminate cancer. In organ transplants, the immune system may attack the new organ as "foreign" unless physicians can find ways to suppress the response. And many scientists believe that regulating and stimulating immune system molecules like interferon provides the best hope for developing cures for cancer.

CHAPTER 16

THE CODE OF LIFE

When Jim's wife, Jeanne Waples, returns to a café in her home town of Red Lodge, Montana (population 1,100), she has no trouble picking out members of the town's various families, even if she's never met them before. "You must be Joki," she'll say to a blond, Finnish-looking boy. She's not a great detective, but living in a small town can be an exercise in practical genetics. Members of families do tend to look like one another, and with a little experience anybody can pick out resemblances.

But the genetic code is more than skin deep and it bears upon much more than casual questions of physical appearance. When Margee Hindle and Bob Hazen got married they knew that their children would have a fifty-fifty chance of suffering from Lynch's syndrome, a genetic disease that invariably leads to colorectal cancer. Is Margee carrying the gene that afflicted her father and grandmother? Was that gene passed on, like a ticking bomb, to Bob and Margee's children?

We pretty much take for granted one of life's miracles: like always begets like. Bacteria beget bacteria, birds beget birds, bananas beget bananas. Offspring display many traits—both good and bad—of their parents. Each new organism begins with a single cell, yet within that microcosm lies all the information needed to create the whole organism in all its complexity. In every form of life a few different atoms and molecules, in cells with the same kinds of architecture, adopt very different designs. How are such complex and varied blueprints passed from one generation to the next? How are they read? Scientists now realize that every living thing on earth uses the same strategy:

All life is based on the same genetic code.

MENDEL'S PEAS

Genetics, the branch of science devoted to studying how traits are passed from parents to their offspring, didn't begin in a high-tech lab with fancy machines and technicians in white coats. Gregor Mendel (1822–84), an Austrian monk whose work was largely ignored in his own lifetime, performed the first comprehensive and systematic genetic experiments in a secluded monastery garden where he cultivated peas.

Mendel noticed that certain strains of peas bred true: tall plants, if bred together, gave rise to tall offspring, while short parents always yielded short offspring. When he produced a hybrid by fertilizing short plants with pollen from tall ones, the offspring were all tall, but if he then bred those hybrid offspring with one another, three quarters of their offspring came out tall and one quarter came out short. Crossbreeding tall and short plants always resulted in tall and short plants—never medium-sized ones, as one might at first expect.

To explain many years' worth of data of this type, Mendel introduced something he called the "gene," defined as the basic unit of heredity. His idea was that each adult possessed two sets of genes, one contributed by each parent. The interplay between these genes then determined the offspring's characteristics. In this game, however, no compromise was possible—one gene or the other won.

To express this kind of competition, Mendel characterized genes as either dominant or recessive. A dominant gene is one that wins the competition if paired with a different gene. For example, in Mendel's pea plants the gene for tallness is dominant. In the first generation of hybrids, where each offspring has one tall and one short parent, each offspring gets one gene for tallness, one for shortness. The fact that all offspring are tall says that in this situation tallness wins, so this gene must be dominant.

The role of recessive genes becomes obvious only in the second generation. All of the tall first-generation plants carry one gene for shortness, even though this gene is not "expressed" (to use the biologist's term). Nevertheless, the gene is there and can be passed on to the next generation. Each parent, in fact, has a fifty-fifty chance of passing on the recessive gene for shortness, and an equal chance of passing on the dominant gene for tallness.

On the average, then, one fourth of the second generation offspring receive genes for tallness from both parents. These offspring will be tall. Half the offspring receive one gene for tallness, one for shortness. Because tallness is dominant, these offspring will also be tall. The final fourth of the offspring receive a gene for shortness from each parent. These offspring will be short.

The existence of recessive genes explains many well-known phenomena in human heredity—the redhead who crops up in a dark family, for example, or the prevalence of hemophilia in the inbred royal families of Europe in the nineteenth century. The gene for light hair is recessive in humans, and so can be carried along by several generations without being expressed. When two dark-haired parents each carry the recessive gene, however, one fourth of their offspring (on the average) will be light-haired. Similarly, the gene for hemophilia (a condition in which the blood is unable to clot) is recessive in humans, but if families where the gene exists intermarry, the chances of an unhappy double recessive increase.

Millions of families across America know similar uncertainties, for there are hundreds of genetic diseases. Children with Tourette's syndrome may suddenly display violent antisocial behavior. Individuals with retinitis pigmentosa suffer increasing loss of vision as the light-sensing parts of their eyes deteriorate. Each of these afflictions is passed on from parent to child, in nature's cruel game of Russian roulette.

When Mendel introduced the gene it was purely a concept—an idea. Genes had no physical reality, and no one knew what they might be. Today we know that genes are a coded sequence of smaller molecules arrayed along a segment of a much larger molecule called DNA.

In passing from the gene as an idea to the gene as a real thing, we pass from classic Mendelian genetics to modern molecular genetics. This is but one example of what is probably the most important development in the history of biology—the shift in emphasis from studying organisms (like plants and animals) to studying the chemical basis shared by all living things.

DNA AND RNA: MESSENGERS OF THE CODE

It has become a matter of folklore that DNA is the molecule that governs heredity, and that it is shaped like a double helix. DNA is one example of a nucleic acid—the "NA" of DNA (so named because it is found in the nucleus of the cell). Like other important molecules of life, DNA is modular—built up from repeated stackings of simple building blocks. In the case of DNA and RNA, the two types of molecule that carry the genetic code, there is modularity within modularity: DNA consists of a string of basic building blocks called nucleotides, which are themselves built from smaller molecules. Think of the process of making DNA as analogous to putting together a book from smaller components like letters, words, and sentences. Assembling DNA involves fitting the subassemblies together rather than starting everything from scratch.

Letters of the Code

Nucleotides represent the "letters" of your genetic code. Your body contains untold trillions of these molecules, like a printer's shop overflowing with type just waiting to be set into prose. Each nucleotide is formed from three smaller molecules. The simplest of these, the phosphate group, consists of a phosphorus atom surrounded by four oxygens. Next comes a sugar: deoxyribose in DNA (the "D" of DNA), and ribose in RNA, another important nucleic acid. The phosphate and sugar are linked to one of four different molecules referred to collectively as bases. The four base molecules—adenine, cytosine, guanine, and thymine—are similar in size but quite different in shape, like four different letters in the alphabet. In fact, they are usually represented by the letters A, C, G, and T.

Each nucleotide is an L-shaped combination of one sugar, one phosphate, and one base. Taken alone, this molecule is not very interesting, just as a page with only one letter makes pretty dull reading. But link nucleotides together in just the right way and you have created the book of life.

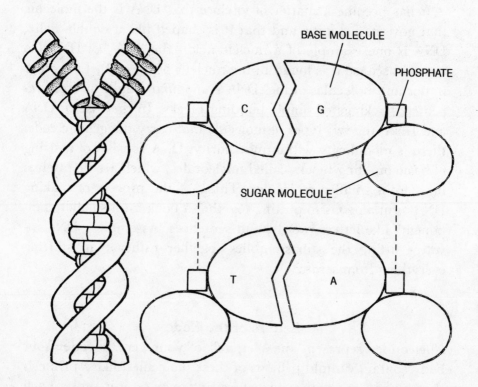

The double helix DNA molecule, which carries the coded blueprints for all living things, is like a twisted ladder with sugar molecules for the vertical sides and base molecules for the rungs.

The Double Helix

The best way to think of the DNA structure is to imagine building a ladder out of nucleotides. The sugar-phosphate part of the nucleotide forms the sides of the ladder; the base pairs hook together

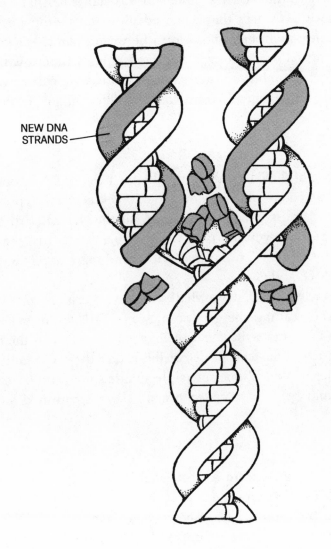

NEW DNA
STRANDS

DNA replicates itself by first splitting down the middle like a zipper. Each half of the original double helix attracts complementary bases to form two new identical molecules.

to form the rungs. Once you have built the ladder in this way, imagine taking the ladder's top and bottom and twisting in opposite directions. What you have is the famous double helix of DNA.

The sequence of the ladder's "rungs" is crucial. The shape of the base molecules is such that when adenine and thymine are brought together, hydrogen bonds form between them and they lock together into a solid "rung." The same thing happens with guanine and cytosine, but not with any other possible pairing of the four bases. Thus, there are only four possible "rungs" to the ladder:

$$AT \qquad TA \qquad GC \qquad CG$$

The sequence of the bases along the double helix of DNA contains the genetic code—all the information a cell needs to reproduce itself and run its chemical factories, all the characteristics and quirks that make you unique. All of the genetic words, sentences, and paragraphs are spelled with combinations of four "letters": A, T, G, and C.

RNA (ribonucleic acid), as we shall see, plays a critical role in transferring and reading the genetic messages. RNA molecules are similar to DNA molecules, except that (1) the sugar in their nucleotides is ribose instead of deoxyribose, (2) they have only half the "ladder"—that is, one sugar-phosphate spine with bases sticking out, and (3) a base called uracil (U) is substituted for thymine.

The Genetic Xerox Machines

All living systems must reproduce to survive, so at the most fundamental level there must be a way to copy molecules of DNA. The procedure for doing so is simple. First, enzymes "unzip" part of the DNA, breaking the bonds that hold the base pairs together (think of the enzyme as sawing through the rungs of the ladder). The unzipping opens the unattached bases to the environment. Inside the nucleus of cells, free-floating nucleotides are attracted (and become attached) to the exposed bases.

For example, suppose a particular rung of DNA is composed of the bases C and G. When the rung splits, the newly freed C will attract a free nucleotide with a G, while the newly freed G will attract a C. So the original CG rung is replaced by two identical CG rungs. Step by step, rung by rung, this process repeats as each of the separated halves plucks a replacement for its missing partner from the surrounding cell fluid. When this has been done for all of the rungs in the ladder, two identical double helix molecules stand where only one stood before. This process guarantees that each generation of cells has DNA identical to the preceding one.

The Genetic Code

The sequence of bases—A, C, G, and T—along the DNA molecule forms a coded message that tells the cell how to manufacture protein molecules. Since proteins serve as enzymes for reactions between molecules in cells, they determine every function the cell carries out: the very biochemical identity of the cell depends on information coded into the DNA.

In order to go from a sequence of base pairs on the DNA to a protein performing its function in the cell, two things must happen. First, the information on the DNA must be read and carried to the place in the cell where the protein is to be built. Then the coded information must be translated into the specific sequence of amino acids that results in the desired protein. These two functions are carried out by two different kinds of RNA molecules.

The transcription of the information on the DNA molecule occurs by a process that resembles the copying of the DNA molecule we've just described. An enzyme "unzips" a stretch of DNA, and nucleotides that are normally floating free in the nucleus link onto the exposed bases to form a molecule of RNA. If there is a sequence TGC along the DNA, for example, the corresponding sequence along the RNA will be ACG. In this way, the entire sequence of bases in the stretch of DNA is copied onto the smaller molecule of RNA, much as a photograph is copied onto a negative.

The type of RNA made by this transcription process is called "messenger" RNA, or mRNA for short. These molecules are much smaller than the DNA (after all, they copy only short segments of the total). They can therefore pass out of the nucleus through small openings, in the nuclear membrane and move into the main body of the cell, carrying the coded information as they go.

When the mRNA reaches the place in the cell where its work is to be done, a second form of RNA called "transfer" RNA (tRNA for short) comes into play. This key-shaped molecule has three bases along its top and a site on the tail that attracts an amino acid. There are four different kinds of bases, so there must be 64 ($4 \times 4 \times 4$) possible arrangements of three bases, and thus 64 different varieties of tRNA, equivalent to 64 different three-letter words in the code of life. Each tRNA has one triplet of bases along the top and attaches to one and only one type of amino acid along the bottom. Since there are 64 possible triplets and only 20 amino acids, the genetic code contains redundancies—several different triplets can produce the same amino acid, like different words that mean the same thing.

Transfer RNA works as shown in the illustration. Bases along the top of the molecule are attracted to their counterparts along the mRNA, and line up accordingly. When the bases are aligned, so too are the amino acids at the other end of the tRNA. These amino acids, held in their proper sequence by the RNA molecules, then link together to form the protein.

The term "genetic code" refers to the link that leads from three base pairs on a DNA molecule to a particular amino acid holding a certain position along a protein chain. This is not an abstract concept. If you reflect on the fact that everything chemical in your body is manufactured, moved, modified, and used by proteins that are built from scratch from a sequence of aminio acids precisely dictated by the DNA sequence, you will see why we say that DNA holds the secret of life.

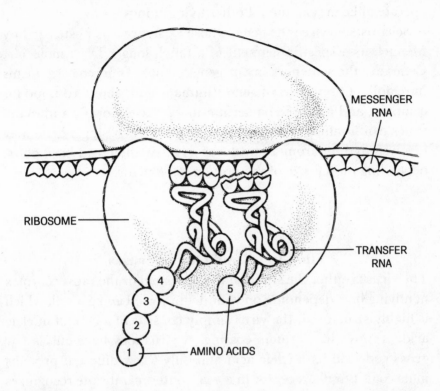

Each key-shaped transfer RNA molecule has three bases along the top that link to the appropriate bases on a strand of messenger RNA. At the bottom of each transfer RNA is a specific amino acid. As transfer RNA molecules line up at one end along messenger RNA, the amino acids line up to form a protein at the other end.

The Universal Code

Just as a simple alphabet can be used to form a large number of words, a simple genetic code can produce a wide variety of living things. You are different from a blade of grass or an orangutan because the sequence of bases in your DNA is different from the sequence in theirs. The basic processes described above, however, operate alike in you and all other living things.

Scientists now see the gene, whose existence was postulated by Mendel, as a specific segment of a much longer DNA molecule. Genes are the sentences in our genetic book. Different organisms have different numbers of genes (humans have about 100,000 for example, and a simple bacterium might have 4,000), and many genes can fit along a single molecule of DNA. A single gene may involve anything from a few dozen to a few thousand base pairs, depending on the size of the protein to be made.

Flu "Bugs" and Other Viruses

The virus is either the simplest living system or the most complex nonliving one, depending on your definitions. Unlike a cell, which is highly structured, the virus simply consists of a core of nucleic acid wrapped in a protein coating. A virus may have only a few genes coded in its nucleic acid and only a few different proteins in its coat. But if a receptor in a cell's outer membrane recognizes one of those outer proteins, the virus can fool the cell into opening the door.

Once inside, the virus appropriates the cell's machinery to produce more viruses. The nucleic acid in most viruses is RNA coded to reproduce itself and its proteins. Once released inside the cell, this RNA pushes the cell's own mRNA aside and starts to direct the synthesis machinery. When the resources of the cell have been plundered to produce many new viruses, the cell dies and the viruses are released to repeat the cycle.

Some viruses contain strands of DNA and enzymes that allow those strands to be inserted into the cell's own genetic code. This action can disrupt the smooth workings of a cell, and thus threaten the entire organism. One of these so-called retroviruses is responsible for AIDS.

SEX—A GOOD IDEA

Your cells' genetic machinery is designed to make exact copies of the coded message, so why isn't every human exactly the same? Why don't we all have exactly the same appearance, the same strengths and weaknesses? The answer is sex.

Chromosomes

DNA does not just float around loose in the cell's nucleus. It is stored in structures called chromosomes, which consist of one long double helix of DNA wrapped around a core of proteins. Chromosomes are something like thread around a spool, but with a single strand of DNA being wrapped sequentially around many adjacent small spools. Different species have different numbers of chromosomes: humans weigh in with 46 (23 pairs), while mosquitoes have only 6, and goldfish have 94.

In many situations cells reproduce by themselves—asexually. This happens when you get a cut that heals, when a plant grows, or when pond scum spreads during warm summer months.

The division of a single cell into two identical daughters is called mitosis. It is a complex process, easy to describe but hard to understand in detail. Mitosis begins when the chromosomes replicate themselves, each length of DNA splitting in half and copying itself. The resulting replicate chromosomes join together and look like the letter X when viewed through a microscope.

At this point a network of proteins (called spindles) forms in the cell and the matched pairs of chromosomes split up, one being pulled to each "pole" of the cell. This separation complete, another band of proteins forms around the cell's "equator" and squeezes down, separating it into two halves, each of which has a full complement of chromosomes. The end result of mitosis: two cells, each with identical genetic codes, where only one existed before.

Cells in your body are constantly dividing, and different cells have different cycle times—those in the intestinal walls divide every day, those in the skin every few weeks, for example. Only cells in the nervous system stop dividing at adulthood.

Meiosis — Reproduction with Sex

In humans and other organisms that reproduce sexually, each parent contributes one chromosome to each pair. This simple fact has two important consequences. First, it means that each off-spring is similar to but different from both parents. Second, it means that the pairing of chromosomes provides the mysterious mechanism that Mendel first discovered in his garden of pea plants. Each gene is one sequence of bases along the double helix of DNA that makes up either the mother's or the father's contri-bution to a chromosome pair. The "unit of heredity" is part of a physical molecule transferred from parents to offspring.

While the goal of mitosis is to produce a daughter cell that is identical to the parent, reproduction of organisms (as opposed to cells) has a somewhat different goal. If each parent contributes half of the genes to the offspring and the number of the offspring's genes must remain the same, then there must be a mechanism for producing a daughter cell with half the number of chromo-somes of the original.

The process of cell division that begins with a cell with a full complement of chromosomes and ends with cells with half this complement is called meiosis. Meiosis occurs only in certain spe-cialized cells in the reproductive system. In its initial stages, mei-osis is identical to mitosis — the chromosomes replicate them-selves. But then, instead of splitting up, the pairs themselves are drawn to the "poles" of the cell along the spindles. After this, another set of spindles forms at right angles to the first and the chromosome pairs are separated. The result: the chromosomes are grouped in the four quadrants of the cell, and each group has exactly half the number of chromosomes in an ordinary cell. The cell then divides four ways, producing four cells, each containing one of the chromosome groupings. Thus does meiosis produce the male's sperm and the female's ova.

The first step in the production of a new organism is fertilization, the union of the ovum and sperm of the two parents. We are used to thinking of fertilization in human, or at least mammalian, terms, but the same things happen in plants, where ordinary pollen is the sperm. Fertilization produces a single cell, the zygote, with a full complement of chromosomes—one chromosome in each pair coming from the sperm, the other from the ovum. This process on the cellular level explains Mendel's notion that an offspring receives half of its genetic endowment from the mother, half from the father.

Does Life Begin at Conception?

The question of whether life begins at conception comes up constantly in the debate over abortion rights in the United States. This is not a scientific question, but a legal, moral, and ethical one. It illustrates an important point about what science can and cannot do. Science is very good at answering quantitative questions about how the universe works, but cannot provide answers to some very important questions about how we should behave as individuals or a society.

A scientist can tell you the following facts:

• At the moment of conception, two strands of DNA come together in a combination that never existed before.

• Each of these two DNA strands existed previously in parent organisms and were themselves the result of a unique pairing. This process probably extends back billions of years.

• The new combination is incapable of independent existence for months after conception, and is entirely dependent on the mother during that time.

Whether "life" begins at, before, or after fertilization is not something that a scientist can answer as a scientist, although he or she may well have a deeply held opinion based on philosophical or religious arguments.

New Reproductive Technologies

New technologies offer hope for would-be parents who are unable to conceive. Many couples are infertile because of mechanical defects in their reproductive system. If both parents produce viable sex cells and the mother's uterus is healthy, however, the eggs and sperm can be combined outside of the mother. Fertilization can take place in a test tube—*in vitro* (literally, "inside the glass") in the jargon of biologists. The fertilized egg, a living human embryo, is then implanted in the mother and pregnancy proceeds normally.

As reproductive biologists learn more about how to induce and control fertilization, troublesome ethical and legal questions arise. Several eggs are usually fertilized at once, and most of those embryos are frozen and stored. If a couple divorces, who has the legal rights to frozen embryos? New techniques allow physicians to identify the sex and genetic traits of test tube embryos before implantation. What are justifiable genetic criteria for rejecting a viable embryo? If the biological mother is unable to bear children, a surrogate mother may grow and nurture the embryo. What are the rights of the surrogate mother? Zoo biologists have successfully bred rare species by using surrogate mothers of a different species. Where do we draw the line with human embryos?

FRONTIERS

Genetic Engineering

The double helix structure of DNA was discovered in 1953. In the short span of time since then, an enormous increase in our ability to understand life at the molecular level has taken place. One consequence of this work is that we have now begun to acquire the ability to manipulate DNA molecules in living things to produce fundamental changes in living systems. This ability in turn has given rise to a new industry, loosely called genetic engineering. In our opinion, this burgeoning new technology has the potential to surpass the microelectronics industry, both in terms of insight into the workings of nature and in its effect on our daily lives.

The central operation in genetic engineering is the production of recombinant DNA or "gene-splicing." Although complex in practice, the central idea is simple. The biologist uses a particular enzyme to make a staggered cut in a DNA molecule, leaving several bases free. Another strand of similarly cut DNA is then brought up. If the bases are complementary (e.g., if a free A on one strand matches up with a free T on another), the two lengths of DNA can bind together, or "recombine." The result: a new strand of DNA, containing sets of genes from each of the two original molecules. And if one splice can be made, so can two or more, so the same technique can be used to insert genes from one strand of DNA into another strand.

There are many ways to apply the developing technology of gene splicing. One common technique involves removing DNA molecules from bacteria, inserting a gene, and then reinserting the altered molecule. For example, the human gene that produces insulin can be inserted into a bacterium in this way, and the insulin produced by the altered bacterium and its descendents can be extracted. In fact, we now produce much of the insulin used to treat diabetes in this way—a great improvement over the traditional method of extracting it from the pancreas glands of dead pigs. The cancer-fighting drug interferon is also produced commercially from recombinant DNA, and many more such biologically engineered drugs will probably be produced in the future.

Alternatively, a gene can be spliced into the ovum of a plant or animal to produce new "genetically engineered" offspring. Frost-resistant plants, pigs that gain weight faster, and laboratory mice whose immune systems copy those of human beings already have been produced in this way. The ultimate science fiction nightmare, the genetically engineered human being, remains comfortingly far in the future at this point.

The Human Genome Project

In the coming years you'll see many references to the Human Genome Project, the first "big bucks" enterprise proposed for the biological sciences. The goal of this project is to produce a listing, base pair by base pair, of the entire human genetic code—all 23 chromosome pairs with about three billion base pairs. (A base-by-base survey is called "sequencing," as opposed to "mapping," which simply locates the genes along the DNA.) Microbiologists estimate that the project will take about a decade to complete, at a cost in the region of three to four billion dollars. The genome project provides a key for understanding and perhaps eventually curing hundreds of genetic diseases. Yet even if it does nothing else but stimulate the technology of genetic engineering, it should be well worth the initial cost.

Gene Regulation and Differentiation

Every cell in your body (except for the sex cells) contains exactly the same DNA, and therefore exactly the same genes. Yet not every cell performs the same function. For example, they contain the code for making insulin, but only a relatively small number in the pancreas actually do so. Most genes in a cell are not actually used. The mystery of how genes are turned on and off (or "regulated") remains an area of intense research today.

The problem may be related to another aspect of molecular genetics—the existence of so-called "junk" DNA. Only about 5 percent of all the DNA in your body is actually taken up by your genes. The function of the rest, which is intermingled with the genes in all chromosomes, is unclear. Biologists used to think that it was not used at all, hence the term "junk." Today, people are starting to think that this DNA may actually contain instructions for turning genes on and off. If this is true, "junk" DNA may well turn out to be just as interesting as genes.

Another problem closely associated with that of gene regulation involves the question of how complex organisms develop from a single cell. All the cells in your body arose from a single cell, but all are now very different and could not be turned into one another. Cell differentiation is one of the main concerns of embryology. It appears that DNA doesn't code only for proteins and regulation, but also contains instructions that turn genes on and off depending on how cells are developing elsewhere in an organism. Biologists have just started to scratch the surface of this complex problem, and we will not have a complete understanding of molecular genetics until gene regulation and cell differentiation are understood.

DNA Fingerprinting

"DNA CLUE CLOSES BOOK ON SLAYING"

Headlines like this one, which followed the arrest of a thirty-four-year-old rapist and murderer in the suburbs of Washington, D.C., are becoming commonplace. In this case, police quickly identified a prime suspect based on an abundance of circumstantial evidence, but there were no fingerprints and no witnesses to the crime. Nevertheless, the suspect left behind one piece of damning evidence—his own sperm cells, with their unique genetic pattern locked in DNA.

"Genetic fingerprinting" relies on the assumption that every person's DNA is distinctive. Technicians isolate DNA from blood, saliva, hair follicles, or semen and place it in an enzyme solution that chops the long DNA strands into thousands of shorter bits. The DNA fragments are placed in a gelatin-like material and subjected to a strong electrical field, which separates all the bits according to their sizes and electrical behavior (some fragments move faster than others in the field). The result is a pattern of stripes on the gelation that looks a lot like the bar code on your groceries. Given sufficient detail, every person's genetic "bar code" is as unique as a fingerprint.

Detectives compared the DNA patterns from the suspect's blood and saliva with that of semen found at the crime scene. The two bar codes matched, and according to FBI specialists only one person in 300,000 would be likely to display a similar pattern.

Not all scientists accept DNA evidence as infallible. The laboratory techniques for isolating and analyzing DNA from a few cells are exacting and subject to a variety of experimental errors. Careful monitoring of test procedures is essential; a few well-publicized cases have been lost because of improper methods. Even if tests are done correctly, it is not always easy to calculate the probability that a match is distinctive enough to convict. Nevertheless, the FBI stands by genetic fingerprinting, and has introduced DNA evidence in hundreds of criminal trials.

The same method is commonly applied in paternity suits. Since half of a child's genetic material is inherited from each parent, scientists can often tell with certainty whether two people are closely related.

CHAPTER 17

EVOLUTION

If you or your children went to public school in the United States, chances are that Creationism—the biblical account of human origins—wasn't a part of the science curriculum. This educational decision was not a foregone conclusion. A dedicated group of scientists has been fighting a series of pitched legal battles in courtrooms across the country—in Arkansas, in California, in Texas—to protect science from what many of us see as one of the greatest threats to the science education of America's children.

This chapter will offend some people, but that is nothing new. The theory of evolution has been offending people for more than a century. Two strongly held views about the origin of our planet and its life are in severe disagreement. Biblical Creationists accept on faith the literal Old Testament account of creation. Their beliefs include (1) a young earth, perhaps less than 10,000 years old; (2) catastrophes, especially a worldwide flood, as the origin of the earth's present form, including mountains, canyons, oceans, and continents; and (3) miraculous creation of all living things, including humans, in essentially their modern forms. If you are a Creationist, the Bible—not nature—dictates what you believe. Creationists subordinate observational evidence to doctrine based on their interpretation of sacred texts. The tenets of biblical Creationism are not testable, nor are they subject to dramatic change based on new data. In other words, Creationism is a form of religion.

The testimony of nature—evidence that anyone can observe and interpret—belies Creationist dogma. If the earth is only 10,000 years old, how could the Grand Canyon have been carved a mile deep in solid rock? How could plate tectonics split apart Europe and North America with spreading rates of only a few inches per year? How could radiometric age dating, based on the steady decay of radioactive elements, give ages of hundreds or thousands of millions of years for most rocks? How could seasonally varying deposits of Mississippi River sediments, coral reefs, and deep ocean deposits contain hundreds of thousands of annual layers, all on top of much older rocks? Nature has much to tell us about our origins, if only we listen without prejudice.

The biblical story of creation has great poetic beauty and metaphorical power. The biblical story of creation (religion) and the theory of evolution (science) are different, complementary ways of answering questions about the origins of life and humans. Because of this fundamental difference, we believe that it is inappropriate to incorporate Creationism into any science curriculum.

The scientific theory of evolution has been developed and modified, challenged and tested, over centuries of geological and biological observations. The theory of evolution leads to specific predictions regarding location of fossils, age of rock formations, and genetic similarities of different species. Evolution is testable and, like any scientific theory, subject to change based on new data. The central idea that has emerged from these studies is:

All forms of life evolved by natural selection.

One must distinguish between the *fact* of evolution and any particular *theory* of evolution, a distinction that will be clear if you think about gravity. There have been many theories of gravity, from Newton to Einstein to (perhaps) a fully unified field theory. Any one of these theories may be wrong, incomplete, or incorporated into another. But if you drop an object, it falls, regardless of which theory you believe. That is the fact of gravity.

In the same way, the fossil record, molecular biology, and geological research all buttress the notion that modern complex life on earth evolved out of earlier, simpler forms. This is the fact of evolution. As with gravity, there are many theories of evolution that purport to describe this process. Any of them, starting with Darwin's, may be wrong or incomplete. The correctness or incorrectness of any particular theory, however, doesn't change the fact of evolution, any more than one can question the fact of gravity.

Most scientists agree about one aspect of evolution. Life seems to have arisen in a two-step process. The first stage—chemical evolution—encompasses the origin of life from nonlife. Once life appeared, the second stage—biological evolution—took over.

CHEMICAL EVOLUTION

On a clear winter's night, gazing into the cold depth of the sky, you can face the brute fact that the universe is a cold, hostile, forbidding place, almost completely devoid of havens for living things. That life should evolve at all is a remarkable thing, requiring just the right temperature, pressure, and chemical elements, as well as a source of energy to combine those elements. The early earth provided all those conditions.

The first requirement for the evolution of life as we know it is an ocean, the mixing bowl for the chemicals of life. The early earth had both an abundance of water and temperatures that remained within the rather narrow range of freezing and boiling water. Within a few million years of the earth's solidification, water covered most of the globe's surface.

Evolution has also required an abundant supply of the six "CHNOPS" elements: carbon, hydrogen, nitrogen, oxygen, phosphorus, and sulfur. All of these components were present in the early atmosphere, which was very different from the air we now breathe. The gases that came from volcanoes to form the first atmosphere were primarily ammonia (NH_3), methane (CH_4), carbon dioxide (CO_2), hydrogen (H_2), and water (H_2O). These gases mixed with the wave-tossed surface layers of the early ocean, which thus contained all the essential elements of life.

The Miller-Urey Experiment

It is a big step from an ocean with a few essential chemical elements to a living organism. In 1953 Stanley Miller and Harold Urey at the University of Chicago designed an experiment to find out what natural process might have formed the complex molecules necessary for life.

Miller and Urey tried to reproduce the early earth's environment in a jar. Into the glassware they poured water and created an atmosphere of ammonia, methane, water, and hydrogen gases. They continually heated and mixed the gases and water while electric sparks, simulating lighting, added energy. The results were amazing. Within a few days the water turned brown, and chemical analysis revealed amino acids — the building blocks of proteins.

Subsequent experiments using other combinations of gases or ultraviolet radiation yielded similar results. Amino acids, sugars, and other essential molecules of life formed in every case. The longer the experiment lasted, the more diverse and concentrated the molecular broth. For a time the public feared that some new and dangerous form of life might actually arise from the test tube. In fact the Miller-Urey molecules were several steps removed from life, but they demonstrated that under the right conditions the molecules of life will form in abundance.

Today, there are laboratories all around the world devoted to "origins of life" studies — laboratories where high-tech descendants of the Miller-Urey experiment probe the ability of the early earth to produce ever larger and more complex molecules. This research demonstrates that there is no problem making extremely complex molecules in conditions like those in the atmosphere or oceans of the primitive earth.

Another source of complex molecules on the early earth was the meteorites that were still falling. We know that modern meteorites contain organic molecules like amino acids, and these molecules would have added to those being produced by Miller-Urey processes. A minority of scientists argue that life actually began with spaceborne debris, a proposition most view with skepticism.

The Primordial Soup

One theory of how life began on earth involves the creation of the appropriately named "primordial soup." This theory holds that radiation from the sun and lightning from the sky provided the energy required to combine simple gases into complex carbon-based molecules, which must have crowded the early oceans. For hundreds of millions of years life's chemicals were created and concentrated in the ocean's upper layers. There may have been scores of different amino acids, linking together to make primitive proteins. Lipid molecules clumped together to form membrane-like sheets and spheres. And DNA-like strands of sugars and bases also may have been present from time to time in that pre-life soup.

We do not know how life arose from the primordial soup. This remains the greatest gap in our knowledge of the development of life. Many scientists believe that millions of years of random mixing and shuffling of molecules culminated in the appearance of one living cell—an object that could consume surrounding chemicals to make exact copies of itself. We can't say for sure when that pivotal event occurred, but some of the earth's oldest rocks, about 3.6 billion years old, show evidence of one-celled life. Suddenly it was a whole new ball game on planet Earth.

The problem of how the first cell was created would be less difficult if some modern "origin of life" theorists turned out to be right. Their idea is that although the primoridal soup may well have formed, it was not crucial in the development of life. Instead, lipids made in the early ocean by processes of the Miller-Urey type (a primordial oil slick?) wrapped themselves into little bubbles, and the chemical process of molecule building took place inside those bubbles. In these theories, instead of having the molecules form first and then separating them from the environment, the primitive cell walls form first, and the molecular processes go on inside them. The authors find this an attractive alternative to the currently popular theories of the origin of life. As with all scientific theories, however, only further tests will tell us whether we're on the right track.

BIOLOGICAL EVOLUTION

The first living cell was not threatened by predators and it lived in an ocean filled with nutritious molecules. It had no competition from any other life form. It may have taken hundreds of millions of years to create the first cell, but within a relatively short time, perhaps only a few years, that cell's offspring probably filled the world's oceans, consuming much of the organic raw materials and greatly reducing the chance that any other type of cell would spontaneously arise. In essence, the first cell, once it appeared, preempted other possibilities of life.

Natural Selection

All the diversity of life on earth—trees, mushrooms, amoebas, and humans—evolved from the first cell by the process of natural selection. By the mid-nineteenth century most geologists and paleontologists had accepted the fact that no species lasts forever; new species appear and old species become extinct. Still, the mechanism of these changes was a mystery. Charles Darwin, a meticulous British naturalist, proposed a solution brilliant in both its power and simplicity. Twenty years of research culminated in the 1859 publication of *On the Origin of Species by Means of Natural Selection,* which remains one of the pivotal events in the history of science.

Darwin studied individual variations in domesticated and wild animals and arrived at three major conclusions. First, every species exhibits variations: size, strength, coloration, and hundreds of other traits vary from individual to individual. Second, many traits are passed on from parents to children: taller parents tend to produce taller children and so on. Both of these ideas were second nature to readers in Darwin's native England, where animal breeders had artificially selected desirable traits for centuries.

Darwin's great contribution was the recognition that variations and inheritance of traits also influences survival and breeding in a natural, wild setting. Each offspring, Darwin realized, inherits traits from both parents, yet no child is exactly like either parent.

Most of the variations within a species have little to do with survival—green-eyed flies probably do as well as red-eyed flies—but once in a while a trait matters. If an individual is better able to survive and attract a mate, by virtue of stronger muscles or better camouflage or fancier feathers, chances are that it will pass those traits on to more offspring than its rivals, and eventually the entire population will share them. Darwin called this process "natural selection, or the survival of the fittest."

Variation in a common species of British insect, the peppered moth, provided a striking example of Darwin's view of life. Early in the 1800s light-colored lichens covered many British trees. Light-colored, mottled peppered moths blended in well with this background and so avoided birds and other predators. The industrial revolution, powered by giant coal-burning engines, blackened the skies and tree trunks over much of Britain, eliminating the light peppered moth's protection. As trees became sooty in industrial areas, natural selection favored the dark individuals, who were always in the population, albeit in small numbers up to that point. With the change in bark color, more dark moths survived to produce more dark offspring. By the late 1800s British peppered moths were predominantly dark. The species survived by adapting to the changing environment.

Life is a contest for limited resources. An individual can survive only if it is able to beat the competition and withstand the vagaries of floods, droughts, hot spells, and ice ages. Life uses many different strategies to survive. Most insects like mosquitoes and cockroaches produce vast numbers of young, of which just a few live long enough to reproduce. Most large animals like us, on the other hand, devote a great deal of energy into nurturing just a few offspring. Some species develop remarkable camouflage that mimics bark or foliage or snow, while others sport flamboyant coloration to warn potential predators of poison.

The phrase most often associated with Darwin, "the survival of the fittest," has from time to time been misused to rationalize political or economic actions by a powerful elite. Darwin defined "fit" in a very restricted sense: the most fit individual is the one that passes his or her genes to offspring that can, themselves, produce yet more offspring. Number of descendants is the full measure of biological success.

The Mechanism for Change

The great gap in Darwin's thesis, and an obvious target for his critics, was the absence of any known mechanism to introduce and pass on new traits and variation. Mendel and subsequent geneticists learned part of the story, but not until the structure and function of DNA were determined did the way variations are produced become clear. No matter how reliable the duplication of DNA might be, mistakes do happen. Damage from X rays, ultraviolet radiation, heat, or certain chemicals increases the rate of errors. Over time many small changes, called mutations, creep into a gene. Some errors are unimportant and get passed on to offspring without any discernible effect. Some errors are disastrous and destroy any chance for viable offspring. Some errors lead to genetic diseases, by which some small but critical part of the body's chemical machinery fails to operate properly. And once in a very great while a chance error results in a new, desirable trait that confers an advantage on offspring, and natural selection takes over to spread that trait through subsequent generations. Over the span of millions of years such small changes accumulate to become major differences.

One of Darwin's most profound insights is that life evolves because it is so competitive. Random variations and chance mutations occasionally lead to advantages, which are preserved as *nonrandom* evolution. Giraffes did not evolve long necks by stretching to reach the highest branches. Rather, natural selection favored the animals that by chance happen to be slightly taller. Individual traits vary at random, but nature selects traits by circumstance.

The random mutations acted on by natural selection accumulate and, over long periods of time, produce organisms that differ markedly from their ancestors. This is the basic mechanism by which new species come into existence, although, as we shall see, the exact way that this process occurs is a subject of debate among scientists today. We should point out that normal rates of evolution are easily sufficient to produce all the complexity of life we see around us. Scientists have estimated, for example, that if a population of mice began to change at a rate observed in many organisms today, those mice could be as big as elephants in a few tens of thousands of years!

EVIDENCE FOR EVOLUTION

Research from the earth and life sciences provides copious evidence for the common ancestry and continuous change of life on earth, beginning billions of years ago.

Molecules of Life

The molecular makeup of life provides compelling evidence for evolution from one single cell. All life is built from the same small subset of organic molecules. All of earth's living things, from slime mold to tea roses to humpback whales, have the exact same DNA-based genetic code, with the molecules following right-handed (never left-handed) spirals. Of all the hundreds of different possible amino acids, only twenty different types form all proteins in every organism. It is reasonable to argue that some or all of these chemical oddities arose in the first cell and have been locked in ever since. If more than one cell had arisen independently, then life would surely possess more than one chemical vocabulary.

Cells

The cellular architecture of all life also points to a common ancestry. Every living thing is made of cells, all of which share many of the same chemical and physical structures. There is also an intimate connection between single-celled and multicelled organisms. Even large and complex animals and plants are collections of cells that are often capable of separate existence. Individual human cells are highly specialized to serve as skin or muscle or nerve or organ. But isolate those cells and they revert to single-celled behavior. Single human cells can adopt amoeba-like form, and they feed and duplicate just like bacteria. In one sense humans and other mammals are colonial organisms, formed from trillions of cooperating cells.

Fossils and Evolution

The most dramatic evidence for evolution comes from fossils—the rockbound remains of past life. Fossils form when organisms die and are buried. Mineral-rich waters flowing underground gradually replace the atoms in the organism's hard parts until, at last, we have a fossil—a replica in stone of the original. Ocean floors are littered with shells, scales, teeth, and other durable remains. River valleys and lake bottoms accumulate animal bones and tree trunks, leaves and insects. Fossil relicts appear in sediments from every geological age, but the nature of past life revealed by those fascinating petrified remains changed in striking fashion through the ages. Fossils prove that for almost four billion years of Earth history life has evolved, increasing in both complexity and diversity.

There are, of course, limitations to the fossil record. Most living things do not possess hard parts, so the majority of species are rarely, if ever, preserved in rock. Even preservation of shells and bone is a chancy thing. Most organisms die, decay, and weather away without a trace. The fossils that remain, therefore, are at best a spotty historical record of earth's life.

THE STORY OF EVOLUTION

Geologists and paleontologists divide earth history into five major eras, based on the type of life that dominated the lands and oceans at a particular time. The Archean era (4.5 to 2.5 billion years ago) saw the formation of the solid earth, the filling of ocean basins, and the chemical evolution of single-celled life. The Archean atmosphere, rich in ammonia, methane, and carbon dioxide, was not suitable for life on land, but there is abundant evidence for simple life in the early oceans.

The most ancient ocean sediments contain remains of only single-celled life, commonly microscopic bacteria-like spheres or rods. Such tiny creatures are observed by preparing paper-thin slices of rock, which are studied with microscope. Most of the fossil bacteria are rather nondescript, isolated blobs, but occasionally death caught them in the act of dividing. Other primitive life formed mats of algae, with clearly delineated layers and filaments similar to modern-day Australian algal mats found in ponds and coastal pools. The record of the rock is clear: one-celled life crowded the earth's seas for three billion years.

More complex life and an oxygen-rich atmosphere evolved during the Proterozoic era (2.5 billion to 570 million years ago). The first multicellular plants and animals are found in rocks about one billion years old. Jellyfish, soft-bodied worms, and multicellular algae came to rest in sediments that now form parts of Australia, Europe, and North America. Our knowledge of these ancient flora and fauna is limited because none of these organisms had hard parts. Their preservation required an unusual combination of circumstances: rapid sedimentation, calm waters, and lack of scavenging bacteria.

Life on earth changed dramatically about 570 million years ago, at the start of the Paleozoic era (570 to 245 million years ago), when animals evolved the ability to make hard shells. The fossil record shows a remarkable increase in the diversity of sea life in the space of a few million years. Corals and other colonial animals built vast reef systems near continents. Segmented lobster-like creatures, precursors of modern snails, starfish, sea urchins, and a wide variety of bivalve shells, also abound in ocean sediments from half a billion years ago.

Few of the life-forms in that ancient world would be familiar to today's skin divers, but as the eons passed more and more modern types joined the fossil record. The first jawed fish, land plants, and insects arose perhaps 400 million years ago, while vertebrates crawled from sea to land about 360 million years ago.

Great forests of cycads and ferns developed along with winged insects at the 300-million-year mark, and shortly thereafter large reptiles roamed the surface.

Dinosaurs and other reptiles ruled the land, sea, and air for most of a quarter of a billion years during the Mesozoic era (245 to 65 million years ago). *Tyrannosaurus, Stegosaurus, Triceratops,* and other giant dinosaurs are only the most famous of hundreds of curious beasts that evolved, along with the trees, flowering plants, modern-looking shellfish, and the first small mammals, to inhabit almost every corner of the planet. The heyday of the spectacular reptiles lasted for almost 200 million years, but ended suddenly 65 million years ago for reasons that are not yet fully understood. With the death of the dinosaurs, who up to that time had dominated the battle for resources, mammals were gradually able to evolve, adapt to, and exploit many different environments.

The most recent, Cenozoic era (65 million years ago to present) saw the rise of mammals, which have evolved many diverse forms, including *Homo sapiens.* By 10 million years ago life on earth had a distinctly modern cast. Bats, cats, dogs, and rodents were common. There were a few oddities: elephants were hairy with strangely directed tusks, giant sloths stood as tall as a house, and horses had toes. But birds, fish, insects, and other everyday animals were much like those of today's forests and streams, while the oceans contained a recognizable cast of whales, sharks, and reef life. Still, one prominent moden life-form—the hominids—was missing.

Homo sapiens, the human species, is a remarkably recent product of evolution. There are no human-like fossils older than about 4 million years, just a thousandth of the age of life on earth. The oldest fossil men and women, found in African lake sediments, belong to the genus *Australopithecus* (or "Southern ape"). These remains of the first known mammal to walk upright are believed to be ancestral to our species, and provide a direct evolutionary link between ape-like mammals and humans. The oldest member of the hominid family is represented by the fossil called Lucy—a young woman who died over 3 million years ago. *Australopithecus* disappeared about a million and a half years ago, shortly after our genus *Homo* arrived on the scene. Still, more than a million years would pass before *Homo sapiens* evolved to walk planet Earth—about 200,000 years ago.

The Rate of Evolution

Most of today's evolutionary scientists focus not on whether evolution occurred, but how it occurred. When Darwin first proposed his theory, he argued that evolution proceeds at a slow, steady rate, and that small changes gradually accumulate to produce large ones. This view is known today as "gradualism." More recently, two American paleontologists—Stephen Jay Gould and Niles Eldredge—have proposed an alternative theory that goes under the name of "punctuated equilibrium." In their view, evolution is characterized by long periods of little change, interspersed (punctuated) by short periods of rapid change.

The fossil record simply isn't good enough to allow us to differentiate between these two competing theories. Take trilobites, one of the fanciest fossils of the Paleozoic, which are proving especially well suited to studies of evolutionary rates. Each individual trilobite has a well-defined number of segments, but different species have different numbers. Trilobite experts count segments for many trilobites from different times and identify systematic changes. In some rock sequences changes seem to be sudden, but other deposits reveal more gradual shifts. It is likely that evolution proceeds in both gradual and punctuated ways under different circumstances, but scientists will need more fossil data before they can resolve the issue.

EXTINCTION

The fossil record is unambiguous: life on earth has evolved from one-celled microorganisms to simple soft-bodied animals and plants to the remarkable diversity of forms and functions we see today. Countless millions of species have come into being, and

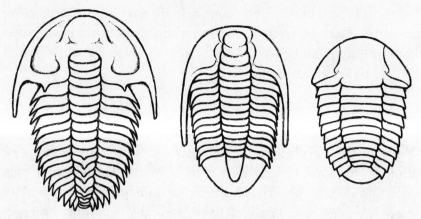

Trilobites, among the most distinctive of Paleozoic fossils, have complex forms that are ideal for studying small evolutionary changes. Some paleontologists spend years collecting trilobites and counting the exact number of body segments or eye facets to document these changes.

almost as many have become extinct. Extinction is as much a part of evolution as the appearance of new forms.

The average lifetime of a species in the fossil record is a few million years. Given the fact that a substantial fossil record extends 570 million years into the past, you can see that the 10 to 40 million species that inhabit our planet today are a small group compared to those that have lived and disappeared in the past.

Mass Extinctions

Darwin and his contemporaries viewed evolution, as well as its inevitable companion extinction, as ongoing properties of life. They thought that species appear and disappear at a relatively constant rate. The fossil record argues against this view.

Life on earth has suffered numerous mass extinctions, the most notable being the one that wiped out the dinosaurs about 65 million years ago. Although we are all familiar with the dinosaurs, two facts about this extinction are less well known than they should be. They are: (1) at the same time the dinosaurs were dying, fully two thirds of all other species on earth were being wiped out in what paleontologists call a mass extinction, and (2) this was neither the most intense nor the most recent mass extinction in the fossil record. At the end of the Paleozoic, 245 million years ago fully 90 percent of existing species became extinct, while another event a mere 11 million years ago had a death toll of approximately 30 percent.

One intriguing idea about the cause of mass extinctions, and one backed by an increasing amount of data, is that they result from the impact of large asteroids or comets on the earth. Such an impact would send an immense, sun-blocking cloud of dust into the upper atmosphere. Months of sunless days would be enough to lower temperatures, deplete food supplies, and kill off most life on earth.

FRONTIERS
Molecular Evolution and Molecular Clocks

If evolution is really driven by the accumulation of mutations in DNA molecules, then one measure of how long it has been since two species shared a common ancestor is the difference in their genetic codes. The study of life's history on earth is now being aided by the molecular biologists, sometimes with surprising results.

Studies of human DNA reveal similarities that seem to indicate that all living humans have a single common ancestor—a woman who lived in Africa about 200,000 years ago and who has been named, appropriately enough, Eve. Similar studies have been used to sort out the human family tree—to show, for example, that we are more closely related to chimpanzees than to the other great apes.

An aspect of molecular studies that has caused a great deal of controversy is the so-called molecular clock. Scientists who propose this idea believe that studies of DNA can not only clarify family relationships, but can tell *when* branchings occurred in the evolutionary tree. Their idea is that changes in DNA occur regularly, like the ticking of a clock, so that if you know how many changes have taken place since two organisms shared a common ancestor, you can tell how long it's been as well.

We expect that there will be a great many molecular contributions to evolutionary theory in the future, and, inevitable turf battles aside, that they will lead to new and exciting insights into our past.

Human Evolution

Although the evolution of humans is not, strictly speaking, more important that the evolution of any other organism, it has an intrinsic interest for members of our species. There are two hot fields of research in this area. One, centered in Africa, concerns finding the oldest human remains and tracing our own family tree to that of the distant ancestors of the apes. This is an intensely competitive field since fossil finds are few and far between, and discoveries receive a lot of media attention.

Another busy field these days is the study of Neanderthal man, our closest cousin on the family tree. Neanderthal became extinct only 35,000 years ago, and most of the debate centers on how like us Neanderthal was—whether he had speech, for example. As with other research in human evolution, our data base is very small so that new discoveries create shock waves every now and then. Everything we know about Neanderthal man is based on the fossilized remains of about a hundred individuals.

A third group of researchers studies the distant ancestors of humans and the great apes, and particularly the question of where our branch of the family tree splits off from the others. This field of study receives less attention in the press than reports of early man, but it is equally fascinating to professionals. As so often happens with the fossil record, there are time gaps that make it very difficult to trace the evolution from the last generally accepted common ancestor to the earliest hominids.

Are Extinctions Periodic?

While some scientists argue about whether mass extinctions are caused by asteroid impacts, others have noted that the many mass extinctions in the fossil record may follow a regularly repeating pattern. It has been suggested that they occur every 26 million years, although there are no good explanations as to what could cause extinctions to occur in such a way.

Work in this area concentrates on two questions: (1) are the extinctions really periodic, or are we being fooled by some subtle statistical effect into thinking that they are? And, (2) what physical mechanism could cause periodic impacts? Theories addressing the second question include the idea of an as yet undiscovered companion star to the sun that jostles the Oort cloud of solar system debris every 26 million years, sending a storm of comets into the inner solar system, and the up-and-down motion of the sun, which takes it through the debris-laden central plane of the galaxy every 26 million years.

CHAPTER 18

ECOSYSTEMS

Lake Victoria, Africa's largest body of fresh water, was once the home of hundreds of species of fish. Among the most important to humans was the tilapia, a delicacy vital to the local economy. Africans harvested and sun-dried tons of the fish, which provided the principal source of protein for millions of lakeshore people.

In the 1960s British sportsmen introduced a new species into the lake—the Nile perch, a voracious predator that grows to several hundred pounds. At first, the tilapia population survived perch predation by escaping to deep water where the perch's visual hunting techniques don't work. But the perch ate other species of fish that limited algae growth. Unchecked, the algae grew out of control, died, sank to the bottom, and decayed, thus destroying oxygen in the tilapia's deep-water sanctuary. With the bottom zone uninhabitable, the unprotected tilapia population is now all but gone. The perch have also eliminated snail-eating fish so snails, which carry dangerous parasites, have become a major health hazard.

The shores of Lake Victoria are nearly equally divided among Kenya, Uganda, and Tanzania. Millions of Africans in hundreds of lakeshore towns and villages are affected by the changing lake ecology. Lake Victoria's native fishermen have switched from tilapia to Nile perch, but the larger fish cannot be sun-dried effectively. The fishermen must roast the perch over wood fires. Now the lake's shoreline is being stripped of trees, resulting in soil erosion and more lake damage. The introduction of one new species has drastically altered an entire ecosystem—an unintended result of man's simple desire for better sport fishing.

This story illustrates a profound truth about living things:

All life is connected.

Living things grow in systems that process the energy and cycle the nutrients needed to support a community of organisms—complex arrangements we call ecosystems. Scientists describe and study ecosystems by mapping the transfer of energy and raw materials (minerals, soil, water) among living things and between living things and their environment.

At the base of the food chain in every ecosystem are self-sustaining organisms—plants and other photosynthetic life. Plants convert energy from the light of the sun to energy-storing molecules, and these molecules serve as the energy source not only for the plants, but for every other living thing. The energy moves up through levels of organisms—those that feed on plants, those that feed on those that feed on plants, and so on—in a complex food web. Eventually, energy leaves the ecosystem and is radiated into space, but the atoms in the molecules remain to be cycled through again and again.

THE HOUSE OF LIFE

Most biologists study life by tackling a small, manageable system—one organ, one cell, or even one molecule. But living systems never occur in isolation. Life requires the complex interaction of many organisms with their surrounding environment. Organisms cooperate and compete, eat or are eaten. Life on earth, along with its nonliving environment, functions as a unit, obeying all the physical and biological principles described in earlier chapters. You have to study the whole, integrated system if you want to understand our planet, and this is where the science of ecology (a word derived from the Greek term for "house") enters the picture. Ecologists study ecosystems, so they concern themselves with all the organisms in a given area and their physical environment.

An ecosystem encompasses no fixed size. Almost any chunk of our planet that includes minerals, air, water, plants, animals, and microorganisms that interact will qualify. An ecosystem could be a swamp, a square yard of meadow, a sand dune, a coral reef, or an aquarium. Natural ecosystems seldom have sharp boundaries: forests merge into fields, shallow water grades into deep water.

Within an ecosystem, each organism fits like a gear in a complex machine. Each organism depends on its fellows, but performs necessary functions for them as well. Termites in a forest depend on trees to produce deadwood, for example, and the trees depend on the termites to clear the ground for new seedlings. The special place occupied by an organism in an ecosystem is called its ecological niche.

All living things on our planet exist in a thin layer at the surface, a layer that extends only a few hundred yards below the solid surface and a few miles into the air. We call this region the biosphere, and it can be thought of as the earth's largest ecosystem.

There is one rule that seems to emerge from studies of ecosystems, a rule that follows from the complexity of the web that connects living and nonliving things. It can be stated simply:

You can't change just one thing.

More grandiloquently, it is:

The Law of Unintended Consequences.

No matter how it's stated, the rule comes down to this: in a complex system it is not always possible to predict what the consequences of any change will be, at least with the present state of knowledge. This means that seemingly small changes in ecosystems can cause large effects, while huge changes might leave the system pretty much as it was.

Having made this point, we should also note that life on earth has survived many wild swings in environment in the past. Nature itself is constantly changing the global environment, so that change in and of itself is not necessarily an evil thing, and it's certainly not "unnatural." Nevertheless, the fact remains that we cannot presently predict with certainty what the ultimate effect of any given change will be.

ENERGY AND THE FOOD WEB

The sun provides the primary source of energy for life on earth. Plants, plankton, and other green life use this radiant energy to convert carbon dioxide and water to energy-rich chemicals—simple carbohydrates—by the process of photosynthesis. Plants harness that chemical energy to produce the more complex molecules—proteins, lipids, and sugars—from which leaves and stems and flowers are made. Plants and other photosynthetic organisms are self-sustaining life. They are the primary producers of energy-storing molecules used by all living things, and scientists refer to them as the first "trophic level" in the environment. It is in the first trophic level that the food chain begins, and it is this level that supplies energy to all living things.

Animals, fungi, and most bacteria can't convert solar energy directly into the molecules they need to sustain themselves, so they seize their energy by eating other life forms. Plants provide the energy source for the primary consumer level of the food chain, which includes grazing animals, caterpillars, and vegetarians, as well as a host of species from bacteria to termites that eat decayed plant matter. Most animals occupy this second trophic level.

Higher up the food chain are animals that feed on other animals in one form or another. Primary carnivores (like wolves) eat herbivores (like rabbits); secondary carnivores (like killer whales) eat primary carnivores (like fish). Other feeding strategies include organisms (like many bacteria, termites, and vultures) that scavenge dead bodies and waste products, and omnivores (like

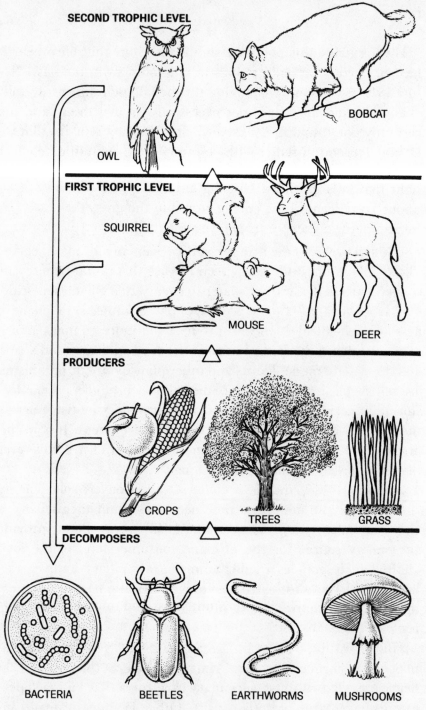

SECOND TROPHIC LEVEL

OWL

BOBCAT

FIRST TROPHIC LEVEL

SQUIRREL

MOUSE

DEER

PRODUCERS

CROPS

TREES

GRASS

DECOMPOSERS

BACTERIA

BEETLES

EARTHWORMS

MUSHROOMS

Energy is obtained by living things through the food chain, which includes several trophic levels. Energy-producing photosynthetic plants provide energy for animal consumers and decomposers.

human beings and raccoons) that get their food from many sources, both plant and animal. All of these animal consumers, however, are ultimately dependent on the photosynthetic producers at the base of the food chain.

The food chain is a very inefficient system. As energy is transferred up from one level to another, much of it is lost at each step. Plants actually use only a few percent of the energy in the sunlight that falls on them. Grazing animals typically recover only about 10 percent of the energy stored in the grass they eat. The lost 90 percent escapes through animal metabolism as heat or is locked in molecules that are not easily digested.

Roughly another 90 percent of the available energy in the food chain is lost in the move from herbivores to carnivores. This ever diminishing flow of energy from lower to higher trophic levels means that each level supports fewer individuals. Vast schools of small fish will feed relatively few big fish. On the African plains there are few lions compared to the large herds of grazing animals. This fact also explains why beef (from the second trophic level) is approximately ten times more expensive than grain, and why there are no lion steaks available in your supermarket.

Each ecosystem is maintained by the energy that flows through it, but no matter how the energy moves through the biosphere, its ultimate fate is always the same. Sooner or later it is converted into heat and radiated back into space, just one more part of the great energy balance that keeps our planet going.

NUTRIENTS AND THE CARBON CYCLE

Unlike energy, which is constantly lost and must be constantly replenished in any ecosystem, the atoms and molecules that make up the structure and nutrients of organisms are recycled. Atoms do not disappear, but move from one organism to another and from one chemical form to another, continuously shifting back and forth between living and nonliving parts of the system. We describe the history of atoms in terms of the so-called chemical cycles. Chemical cycles essential for life include the water cycle, as well as cycles for the elements carbon, nitrogen, oxygen, phosphorus, sulfur, and others.

Each part of the cycle followed by any atom or molecule is complex, with many alternative pathways. Consider the movements of just one carbon atom, entering the cycle from the atmosphere as a molecule of the gas carbon dioxide (CO_2). A blade of grass combines that molecule of carbon dioxide with water through photosynthesis to create part of a glucose molecule. Shortly thereafter, the glucose is processed in cellular chemical factories to form part of a molecule of starch—a basic building block of the cellulose fibers that support each grass blade. The carbon atom has become an integral part of the structure of grass.

A hungry mouse nibbles at the grass, chewing and swallowing the carbon atom, which is added to the mouse's chemical stockpile. The unfortunate mouse is spotted and eaten by an owl, who adds the atom to its own energy reserve. As the owl burns its carbon-rich fuel by respiration, the carbon atom returns to the atmosphere in another molecule of carbon dioxide.

There are many other pathways that carbon might follow. Some carbon atoms end up in the soil as animal droppings or by death and decay. There, bacteria, worms or other scavengers obtain raw materials directly from the carbon-rich earth. Layers of dead plant matter can pile up, become deeply buried, and transform by the earth's temperature and pressure to form fossil fuel deposits such as coal, oil, and natural gas. Snails and beetles convert other carbon atoms into the chemicals that make their hard outer shells. In the ocean, corals and shellfish use a similar process to manufacture durable carbonate reefs and shells, which can accumulate to form thick limestone formations. In the past century humans have altered the natural carbon cycle by burning hundreds of billions of tons of fossil fuel, thus increasing the concentration of carbon dioxide in the atmosphere.

Every other atom essential to life—oxygen, hydrogen, nitrogen, and so on—follows a similar cycle through the biosphere. The details differ from element to element, of course, but the main principle is the same: materials cycle through the biosphere and never leave.

HUMANS AND THE ENVIRONMENT

Humans are an integral part of the ecosystem. Like all other living things, we depend ultimately on the energy in sunlight and the photosynthetic reactions in the first level of the food chain. But humans, unlike any other species, have learned to shape and alter their environment in remarkable ways. By developing agriculture, building cities, and, more recently, building manufacturing plants, we have, for better or worse, profoundly altered the biosphere. Many of the most important questions now on the national and international political agendas have to do with this fact. None of these issues is a purely scientific one; many economic and social factors impinge on each one. Nevertheless, each has a strong scientific component, and it is impossible to discuss any of them intelligently without some basic understanding of the underlying science. Below are three of the many environmental problems about which we all will have to make intelligent decisions as a citizen: ozone depletion, acid rain, and the greenhouse effect.

Ozone Depletion

The ozone molecule consists of three (as opposed to the usual two) oxygen atoms. Only about one molecule per million in the atmosphere is ozone, but these molecules play a crucial role in our environment in two ways. Ozone near the surface of the earth ("bad ozone") is a noxious pollutant, irritating to eyes and lungs. Ozone 50,000 feet up, on the other hand ("good ozone"), absorbs the sun's harmful ultraviolet radiation and thus provides an effective sun screen for those of us living on the ground. Without the ozone layer, humans and other terrestrial life would be constantly bombarded with high-energy radiation and consequently put at risk of medical problems such as skin cancer and eye damage.

The ozone layer is at risk today because of the widespread use of a class of chemicals known as chlorofluorocarbons (CFC's for short). CFC's are used extensively as the working fluids in refrigerators and air conditioners, as cleaners during the manufacture of microchips, and in the manufacture of foam products. When they were first used extensively in the 1960s, the molecules' stability was considered an asset since they wouldn't break down and add to pollution. But that very stability has led to problems, because CFC's last long enough to filter through to the upper atmosphere. There the molecules' chlorine atoms act as catalysts in a complex set of reactions that convert two molecules of ozone to three molecules of ordinary oxygen, depleting the ozone layer faster than it can be recharged by natural processes—another example of the Law of Unintended Consequences.

In 1984, scientists working in the Antarctic made a startling discovery that focused world attention on the ozone layer. During the months of September and October (the Antarctic spring), the concentrations of ozone above the pole dropped by 50 percent. This celebrated "ozone hole" has reappeared, with varying degrees of intensity, every year since. It appears that the massive ozone depletion associated with the hole is the result of the special conditions that are found in Antarctica—the isolation of the air during the Antarctic winter and the presence of ice clouds that form during the period when the sun doesn't shine. Scientists can't predict whether the present Antarctic trends will be repeated in more populated latitudes. Some researchers have reported instances of ozone depletion over nearly the entire earth.

The public has been justifiably concerned about the progressive destruction of the ozone layer, and a sense of urgency has prompted measures to reduce the use of CFC's.

Chemical industries have been quick to respond to the evident danger—Du Pont, for example, has announced plans to phase out all production of the chemicals. There is an agreement among industrial nations to cut the manufacture of CFC's by 50 percent by the year 2000, the worldwide production of CFC's has already been greatly reduced, and a massive search is on for safer chemical substitutes.

Some commentators feel that the response to the ozone problem has been too precipitous. There is, for example, evidence that actual levels of ultraviolet radiation at the surface have not increased as the ozone levels have dropped (perhaps because of the growth of "bad ozone" lower down), and few people engage in sunbathing during the Antarctic spring. Nevertheless, restrictions on CFC's (and perhaps even an outright ban) are prudent responses to our present situation.

As environmental problems go, ozone depletion is a relatively simple one to deal with. The solution is obvious, its cost relatively low, and it requires no real change in behavior or lifestyle to reverse the present trend.

Acid Rain

The burning of fossil fuels like coal and oil produces a number of waste products. Some of these, such as carbon dioxide and water, are inevitable, a result of the chemical reaction that releases energy when oxygen from the air combines with atoms in large hydrocarbon molecules. Other waste products, however, arise because common fuels like coal and petroleum products always include more than just carbon and hydrogen in their makeup. In fact, the frequent presence of nitrogen and sulfur atoms creates the acid rain problem.

When gasoline or coal burns, oxygen combines with nitrogen and sulfur atoms, and the molecules formed in this way enter the atmosphere. There, further chemical reactions take place that produce molecules of nitric and sulfuric acid (the latter is the acid in your car's battery). These molecules are incorporated into raindrops, creating a very dilute acid. When the water falls to the surface of the earth, we call it acid rain.

When intensely acid rain falls in natural settings, it can produce many problems. Lakes become more acid, causing fish populations to dwindle or disappear. In some cases, living trees can be stunted or killed. In Europe, a new word, *Waldsterben* ("forest death"), has been coined to describe this phenomenon. In cities, acid rain eats away at old buildings, irretrievably damaging ancient monuments. This effect is particularly pronounced when the buildings (like Westminster Abbey in London) are made from limestone.

Acid rain is a regional rather than a global problem. The acid rain in New England and southern Canada originates in the smokestack industries of the Ohio Valley, while acid rain in Scandinavia comes from the industrial region of central Germany. In the 1950s and '60s, factories in the Midwestern United States were equipped with tall smokestacks to reduce local air pollution. They were successful in doing so, but, in a perfect example of the Law of Unintended Consequences, got rid of the pollutants by injecting them into airstreams on their way to New England, where they now appear as acid rain.

We know how to reduce or eliminate acid rain. Either the sulfur and nitrogen compounds have to be removed at the source or cleaner fuels have to be substituted for coal and oil. The first strategy employs smokestack scrubbers and automobile catalytic converters. The second uses nuclear reactors in generating plants, and methane or alcohol fuels in cars.

We would place acid rain in an intermediate level on the scale of environmental problems facing the nation. The solution to the problem, while expensive, is fairly easy to specify and well within the means of the industrialized nations.

The Greenhouse Effect

Although it is possible to talk about removing materials that cause acid rain from smokestacks or exhaust pipes, or even converting to fuels that do not produce those materials, one product must inevitably be produced whenever we burn a fossil fuel: carbon dioxide. Whenever you drive a car, cook food, or use an electric light, chances are that you are adding carbon dioxide to the atmosphere.

This addition of carbon dioxide to the atmosphere gives rise to what scientists call the greenhouse effect. In a greenhouse (or even a car left in the open with the windows rolled up), sunlight passes through the glass and is absorbed by materials on the inside. The heated material then gives the energy back in the form of infrared radiation, but the glass is opaque at infrared wavelengths, so the energy remains trapped, warming the interior of the greenhouse or car. Like glass, carbon dioxide transmits visible light coming in from the sun, but absorbs infrared radiation that rises from the ground and holds this heat in the atmosphere instead of reflecting it back into space. The term "greenhouse effect" as applied to the earth refers to the possibility of global warming due to the accumulation of carbon dioxide from the massive burning of fossil fuels that has taken place since the beginning of the industrial revolution.

Several points should be made about the greenhouse effect. First, there has always been carbon dioxide in the atmosphere, so we are not introducing a totally new substance to the environment. In fact, without the partial greenhouse effect from naturally occurring carbon dioxide, the temperature of the earth's surface would be about 20 degrees below zero. Second, carbon dioxide is only one of several infrared-absorbing "greenhouse gases" that humans produce, so the problem is not confined to the consequences of burning fossil fuels. Common substances like methane (natural gas) and even water vapor can add to the greenhouse effect. Nevertheless, carbon dioxide does seem to be the main greenhouse problem.

Finally, an enormous debate is raging within the scientific community about almost every aspect of the greenhouse effect, and many of the extreme statements you hear should be interpreted in terms of that debate. Scientists do agree on two facts: carbon dioxide is a greenhouse gas, and the amount of that gas is increasing. But there is intense disagreement among scientists over two crucial aspects of the greenhouse effect: (1) the question of whether greenhouse warming has already started, and (2) the question of how much warming there will eventually be.

The first question is complicated by the fact that the weather normally fluctuates a great deal from year to year, and a few abnormally hot (or abnormally cold) years do not necessarily indicate the beginning of a trend. Furthermore, there is widespread debate among meteorologists about the reliability of weather records. One example of the kind of problem faced by those trying to establish a warming trend: the most reliable long-term weather stations tend to be located at major airports, and the areas surrounding these stations have become highly urbanized over the last two decades. If such a station records a one-degree warming "trend," are we to blame it on greenhouse warming or on all the newly poured concrete in the area?

To address the question of the ultimate effects of the addition of carbon dioxide to the atmosphere, scientists use giant computer codes called general circulation models (GCM). Typically, they put a model earth whose atmosphere has twice the normal amount of carbon dioxide in it and then let the computers run for several days or weeks. The result: a prediction of average global warming. Until recently, most models agreed on global warmings

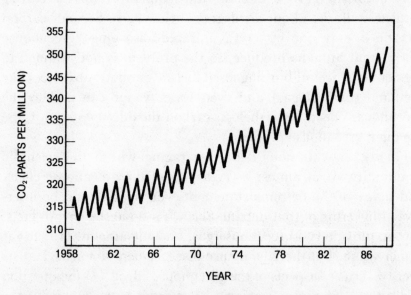

The carbon dioxide content of the atmosphere since 1958 shows a yearly fluctuation (related to the growth cycle of plants) superimposed on an overall 12 percent increase. The overall increase (which may amount to 25 percent over pre-industrial revolution levels) leads to worries about the greenhouse effect.

in the neighborhood of 1.5 to 5 degrees Celsius—an amount not unprecedented in the earth's history and roughly the amount of warming that has occurred since the end of the last ice age. What is unprecedented is the speed of change, about ten times higher than that at any known period in the past. Scientists worry about whether the biosphere can adjust to such a rapid warming.

There are serious objections, however, to the predictions of the GCM. The most serious of these have to do with the role of clouds in the global climate. For technical reasons, the computers break the surface of the earth up into squares about 300 miles on a side—far too large to deal realistically with clouds, which typically are miles to tens of miles across. Since clouds can reflect sunlight back into space before it ever gets to the earth, treating them unrealistically could make a model's predictions useless. In 1989, for example, scientists at the British Meteorological Office managed to put together a GCM that treated some aspect of cloud formation more realistically and found that the predicted total warming would be only about 2 (rather than 5) degrees.

In our view, the greenhouse effect is the most difficult and potentially damaging environmental problem now confronting the human race. The difficulty arises because of the uncertainties (outlined above) in our knowledge of the basic causes and effects needed to evaluate both the magnitude and the probability of the threat. It is further exacerbated by the fact that countering the greenhouse threat, if necessary, will require a massive and unprecedented effort on the part of both industrialized and developing nations to reduce the burning of fossil fuels. Estimates of the cost of reducing carbon dioxide emissions in the United States alone run to several *trillion* dollars. To undertake such efforts in the face of ambiguous and contradictory scientific evidence is not easy in the politicial world.

If it becomes clear that the danger is real, steps like the following would become necessary:

• Stop the rapid destruction of the rain forests in the Amazon. Living trees pull carbon out of the air and store it in their tissues, thereby helping to cancel the effect of burning fossil fuels. Dead trees return carbon to the atmosphere, exacerbating the problem.

- Plant new forests to reduce the carbon in the atmosphere. Calculations indicate that the size of the forest needed to counteract the greenhouse effect falls somewhere between that of the state of Pennsylvania and that of the United States west of the Mississippi.

- Consume smaller amounts of fossil fuels. Policies suggested in this area range from the development of fuel-efficient cars and buildings (the so-called conservation approach) to a complete restructuring of industrial society to eliminate the dependence on gasoline-powered automobiles and develop safe and affordable solar and nuclear power for the generation of energy.

We believe that research in this area should be vigorously pursued so that the risks and consequences of the greenhouse effect can be clearly defined. In the meantime, simple prudence dictates that some steps like those outlined above should be taken, more or less in the same spirit in which people take out insurance policies. After all, planting a tree is a good thing to do, greenhouse effect or no.

Disposable Diapers

Sometimes, in the interest of ecological purity, certain questionable causes become trendy among educated Americans. The great disposable diaper controversy is an example of our failure to understand the way the earth's ecosystem works.

The controversy blew up when it was discovered that plastic disposable diapers (as well as other plastic goods) are not degrading, but are clogging up landfills around the country. Shocked by this example of Man the Polluter, environmentalists around the country called for the development of biodegradable plastic. At the risk of incurring the anger of the purveyors of conventional wisdom, we would like you to ask yourself two questions: where did the material in the plastic come from? And where will it go if it's not in a landfill?

The answer to the first question is easy: virtually all plastics are made from petroleum feedstocks. The carbon in the plastic, then, was taken from underneath the earth, where it started millions of years ago in a prehistoric swamp. The answer to the second question should also be obvious: if the carbon doesn't stay in the landfill, it will return to the air in the form of carbon dioxide *and add to the greenhouse effect.*

It may be, in other words, that ecological consciousness in America is leading us to another exercise in the Law of Unintended Consequences. Like their polluting industrial forebears, biodegradable plastic advocates are forgetting that atoms cycle in the biosphere—they don't go away just because we can't see them. Better to think in terms of using less plastic or recycling what we have, than to take carbon from the ground and guarantee that it will enter the atmosphere.

Nuclear Waste

The good news about nuclear reactors is that they are capable of generating electricity without adding carbon dioxide to the atmosphere, and hence do not add to the greenhouse effect. The bad news is that they pose their own set of potential environmental hazards, primarily those connected with the disposal of radioactive materials once the fissionable fuels in them have been "burned."

The technology of waste disposal has been the center of a great deal of research and development over the past twenty years, and has come far since the early "bury it and forget it" days of atomic energy. Modern techniques for nuclear waste disposal center on the fact that the chemical properties of an atom are independent of whether or not the nucleus is radioactive.

Waste disposal starts with a period of storage of ten or more years, during which nuclei with short half-lives decay. Those with long half-lives are then incorporated into glasses composed primarily of silicon, boron, and oxygen—the idea being that these stable glasses will hold radioactive atoms in place for thousands of years until the nucleus decays. The radioactive glass is then enclosed in a complex canister involving multiple layers of concrete and stainless steel. The current plan is to bury these canisters in a federal underground depository, although the political battle involving the location of that depository continues to rage.

The main worry is that at some time in the future underground water will come in contact with the radioactive materials and return them to the biosphere. Given the time it would take water to dissolve the glass (not to mention the steel and concrete), we feel that this scheme represents the best insurance against future contamination that one can have in an imperfect world.

THE ROLE OF SCIENCE

Science is a way of learning about the cosmos and our place in it. Through science we discover physical laws of governing matter, energy, forces, and motion—universal laws that apply to every life-form and every world, large or small. We explore the atom and the amazing diversity of materials and properties that arises from these building blocks. We identify forces that bind nuclei and fuel stars—forces that can be harnessed for our benefit or unleashed for our destruction.

The scientific method leads inexorably to far-reaching conclusions about planet Earth and our role in its history. The earth is immensely old, formed like all planets and stars from galactic dust and debris. Ours is a dynamic planet with continents and oceans that have been created and consumed a score of times before the present age. Humans are but one small, recent step in the four-billion-year evolution of life on earth, a process that led from a single living cell to the delightful diversity of organisms we find today. Along the way countless millions of species have come into being, and almost as many have become extinct.

At times we like to think of ourselves as exempt from the laws of nature—as somehow special, protected, and above all other creatures. Yet we cannot alter our place in the cosmos as but one species among millions struggling for energy and nutrients to survive. It would be utter folly to ignore the realities of our place in earth's ecosystem.

That being said, it would also be foolish to deny that humans enjoy a special status on earth. Unlike any other species in the planet's history, we have learned to harness resources and shape our environment. We possess the ability to probe our world with the sciences, to appreciate our world through the arts, and to search for the meaning of our unique role with philosophy and religion.

Humans are a doggedly curious species, and science provides our most powerful means for understanding the physical universe. Science is a great human adventure, with formidable challenges and priceless rewards, unimagined opportunities and unparalleled responsibilities. Science lets us view the world with new eyes, exploring backward in time, looking outward through space, and discovering unity in the workings of the cosmos. Armed with that knowledge we can combat disease, create new materials, and shape our environment in marvelous ways. Science also gives us the means to predict the consequences of our actions and perhaps, with wisdom, to save us from ourselves.

ADDITIONAL READING

Hundreds of excellent nontechnical books and articles about science amplify the principles introduced in this volume. We list some of our favorites, without any effort to include everything, and with all due apologies to the authors of many fine books and essays not cited here.

GENERAL

Richard Feynmann, R. Leighton, and M. Sands: *The Feynmann Lectures in Physics*. Cambridge, Mass. Addison-Wesley, 1965.

Rom Harre: *Great Scientific Experiments*. New York: Oxford University Press, 1981.

Alexander Hellemans and Bryan Bunch. *The Timetables of Science*. New York: Simon & Schuster, 1988.

Paul G. Hewitt: *Conceptual Physics*. Boston: Little, Brown, 1974.

George F. Kneller: *Science As a Human Endeavor*. New York: Columbia University Press, 1978.

Stephen F. Mason: *A History of the Sciences*. New York: Collier Books, 1962.

Philip Morrison and Phylis Morrison. *The Ring of Truth*. New York: Random House, 1987.

George A. Sarton: *A History of Science*. Cambridge, Mass.: Harvard University Press, 1959.

Chapter 1: Knowing

Gale E. Christianson: *In The Presence of the Creator: Isaac Newton and His Times*. New York: Macmillian/Free Press, 1984.

I. Bernard Cohen: *The Birth of the New Physics*. New York: W. W. Norton, 1985.

James Gleick: *Chaos*. New York: Viking, 1987.

A. R. Hall: *From Galileo to Newton*. New York: Harper & Row, 1963.

Gerald S. Hawkins: *Stonehenge Decoded*. New York: Doubleday, 1965.

Daniel Kevles: *The Physicists*. Cambridge, Mass.: Harvard University Press, 1971.

Chapter 2: Energy

P. W. Atkins: *The Second Law.* New York: Scientific American Library, 1984.

Chapter 3: Electricity and Magnetism

C. W. F. Everitt: *James Clerk Maxwell; Physicist and Natural Philosopher.* New York: Scribner, 1976.

Michael I. Sobel: *Light.* Chicago: Univeristy of Chicago Press, 1989.

Chapters 4 and 5: The Atom and The World of the Quantum

John L. Casti: *Paradigms Lost.* New York: William Morrow, 1989.

Heinze Pagels: *The Cosmic Code.* New York: Simon & Schuster, 1982.

David Mermin: "Is the Moon Really There When Nobody Looks?" *Physics Today,* April 1985, pp. 38–47.

James Trefil: "Quantum Physics Work." *Smithsonian Magazine,* August 1987.

Chapters 6 and 7: Chemical Bonding and Atomic Architecture

Gerald Feinberg: *Solid Clues.* New York: Simon & Schuster, 1985.

Robert M. Hazen: *The Breakthrough: The Race for the Superconductor.* New York: Summit, 1988.

Duncan McKie and Christine McKie: *Crystalline Solids.* New York: Halsted, 1974.

Arno Penzias: *Ideas and Information.* New York: W. W. Norton, 1989.

Randy Simon and Andrew Smith: *Superconductors: Conquering Technology's New Frontier.* New York: Plenum, 1988.

Jeremy Bernstein: *Three Degrees Above Zero.* New York: Scribner's, 1984.

J. E. Gordon: *Structures: Or Why Things Don't Fall Down.* New York: Plenum, 1978.

J. E. Gordon: *The New Science of Strong Materials: Or Why You Don't Fall Through the Floor.* Princeton: Princeton University Press, 1976.

Helen Rossotti: *Why Things Aren't Grey.* Princeton: Princeton University Press, 1987.

Chapters 8 and 9: Nuclear Physics and Particle Physics

Luis Alvarez: *Alvarez: Adventures of a Physicist.* New York: Basic Books, 1987.

Abraham Pais: *Inward Bound.* Oxford: Oxford University Press/Clarendon, 1986.

Richard Rhodes: *The Making of the Atomic Bomb.* New York: Simon & Schuster, 1986.

Michael Riordan: *The Hunting of the Quark.* New York: Simon & Schuster, 1987.

James Trefil: *From Atoms to Quarks.* New York: Scribner, 1980.

Chapters 10 and 11: Astronomy and The Cosmos

Eric Chaisson: *Cosmic Dawn.* Mass: Little, Brown, 1981.

Timothy Ferris: *The Red Limit.* New York: William Morrow, 1977.

———.: *Coming of Age in the Milky Way.* New York: William Morrow, 1988.

Morton Grosser: *The Discovery of Neptune*. Cambridge, Mass.: Harvard University Press, 1962.

Stephen W. Hawking: *A Brief History of Time*. New York: Bantam, 1988.

William Graves Hoyt: *Planet X and Pluto*. Tuscon: University of Arizona Press, 1980.

Joseph Silk: *The Big Bang*. New York: W. H. Freeman, 1989.

Walter Sullivan: *Black Holes: The Edge of Space, the End of Time*. Garden City, N.Y.: Doubleday/Anchor, 1977.

James Trefil: *The Moment of Creation*. New York: Scribner, 1983.

————.: *The Dark Side of the Universe*. New York: Scribner, 1988.

————.: *Space Time Infinity*. Washington, D.C.: Smithsonian Books, 1985.

Chapters 13 and 14: The Restless Earth and Earth Cycles

William Glen: *The Road to Jamarillo*. Stanford University Press, 1982.

Stephen Jay Gould: *Time's Arrow, Time's Cycle*. Cambridge, Mass.: Harvard University Press, 1987.

John W. Harrington: *Dance of the Continents*. Los Angeles,: J. P. Tarcher, 1983.

Tjered van Andel: *Tales of an Old Ocean*. New York: W. W. Norton, 1977.

Williard Bascom: *Waves and Beaches*. Doubleday/Garden City, N.Y.: Anchor, 1964.

John McPhee: *Basin and Range*. New York: Farrar, Straus & Giroux, 1981.

————.: *In Suspect Terrain*. New York: Farrar, Straus & Giroux, 1983.

————.: *Rising from the Plains*. New York: Farrar, Straus & Giroux, 1986.

Ronald B. Parker: *Inscrutable Earth*. New York: Scribner, 1984.

James Trefil: *A Scientist at the Seashore,* New York: Scribner, 1984.

Chapter 15: The Ladder of Life

Natalie Angier: *Natural Obsessions: The Search for the Oncogene*. Boston: Houghton Mifflin, 1988.

Larry Gonick and Mark Wheelis: *Cartoon Guide to Genetics*. New York: Barnes & Noble, 1983.

John Gribbon: *In Search of the Double Helix*. New York: Bantam, 1985.

James D. Watson: *The Double Helix*. New York: Atheneum, 1985 (reprint of 1968 edition).

Chapter 17: Evolution

Charles Darwin. *On the Origin of Species: A Facsimile of the First Edition*. Cambridge, Mass.: Harvard University Press, 1975 (reprint of 1859 original).

Donald Johanson and Maitland Edey: *Lucy*. New York: Simon & Schuster, 1981.

Roland Mushat Frye (ed.): *Is God a Creationist?* New York: Scribner, 1983.

Stephen Jay Gould: *The Panda's Thumb*. New York: W. W. Norton, 1982.

Paul Kitcher: *Abusing Science*. Cambridge, Mass.: MIT Press, 1982.

Steven M. Stanley: *Earth and Life Through Time*. New York: W. H. Freeman, 1986.

Chapter 18: Ecosystems

John McPhee: *Control of Nature*. New York: Farrar, Straus & Giroux, 1989.

Jonathan Weiner: *Planet Earth*. New York: Bantam, 1986.

INDEX

Robert M. Hazen is the author of more than one hundred and fifty articles and six books on earth science, materials science, history, and music. In 1982 he received the Mineralogical Society of America Award, in 1986 the Ipatieff Prize of the American Chemical Society, and in 1989 the ASCAP–Deems Taylor Award for his writings. He recently led the Carnegie Institution team that discovered the identities of several record-breaking high-temperature superconductors. Hazen is a research scientist at the Carnegie Institution of Washington's Geophysical Laboratory and Robinson Professor of Earth Science at George Mason University. His books include *The Breakthrough, The Music Men, Wealth Inexhaustible,* and *The Poetry of Geology.*

James Trefil, Robinson Professor of Physics at George Mason University, is author of more than one hundred professional papers, three textbooks, and eleven other books on science. A commentator on National Public Radio and a Fellow of the American Physical Society, he also serves on the Committee on Fundamental Constants and Basic Standards of the National Research Council. A former John Simon Guggenheim Fellow, Trefil won the 1983 AAAS Westinghouse Award for excellence in science writing, and his teaching has been recognized by the Innovation Award of the National University Continuing Education Association/American College Testing Program. His books include *The Moment of Creation, A Scientist at the Seashore, Meditations at 10,000 Feet, The Dark Side of the Universe, Reading the Mind of God,* and, as coauthor, *The Dictionary of Cultural Literacy.*

What You Need to Know

AND WHERE TO FIND IT

WHAT YOU NEED TO KNOW